30/03/15 .

PRAISE FOR INFINITESIMAL

'A gripping and thorough history of the ultimate triumph of the mathematical tool... *Infinitesimal* will inspire you to dig deeper into the implications of the philosophy of mathematics and knowledge.'

New Scientist

'Fascinating... Amir Alexander vividly re-creates a wonderfully strange chapter of scientific history, when fine-grained arguments about the foundations of mathematical analysis were literally matters of life and death... You will never look at calculus the same way again.'

Jordan Ellenberg, Professor of Mathematics, University of Wisconsin–Madison, and author of *How Not to Be Wrong*

'A gripping account of the power of a mathematical idea to change the world. Amir Alexander writes with elegance and verve about how passion, politics, and the pursuit of knowledge collided in the arena of mathematics to shape the face of modernity. A page-turner full of fascinating stories about remarkable individuals and ideas, *Infinitesimal* will help you understand the world at a deeper level.'

Edward Frenkel, Professor of Mathematics, University of California, Berkeley, and author of *Love and Math*

'You may find it hard to believe that illustrious mathematicians, philosophers, and religious thinkers would engage in a bitter dispute over infinitely small quantities. Yet this is precisely what happened in the seventeenth century. In *Infinitesimal*, Amir Alexander puts this fascinating battle in historical and intellectual context.'

Mario Livio, astrophysicist, Space Telescope Science Institute, and author of *Brilliant Blunders*

'In *Infinitesimal*, Amir Alexander offers a new reading of the beginning of the modern period in which mathematics plays a starring role. He brings to life the protagonists of the battle over infinitesimals as if they were our contemporaries, while preserving historical authenticity. The result is a seamless synthesis of cultural history and storytelling in which mathematical concepts and personalities emerge in parallel. The history of mathematics has rarely been so readable.'

Michael Harris, Professor of Mathematics, Columbia University and Université Paris Diderot

'We thought we knew the whole story: Copernicus, Galileo, the sun in the center, the Church rushing to condemn. Now this remarkable book puts the deeply subversive doctrine of atomism and its accompanying mathematics at the heart of modern science.'

Margaret C. Jacob, Distinguished Professor of History, University of California, Los Angeles

ALSO BY AMIR ALEXANDER

*Duel at Dawn: Heroes, Martyrs, and
the Rise of Modern Mathematics*

*Geometrical Landscapes: The Voyages of Discovery and
the Transformation of Mathematical Practice*

INFINITESIMAL

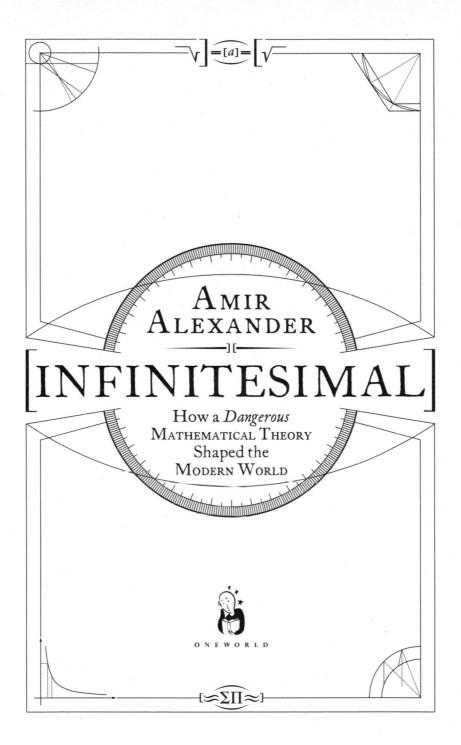

Amir
Alexander

[INFINITESIMAL]

How a *Dangerous*
Mathematical Theory
Shaped the
Modern World

ONEWORLD

A Oneworld Book

First published in Great Britain by Oneworld Publications, 2014
Published by arrangement with Scientific American, an imprint of Farrar, Straus
and Giroux

ISBN 978-1-78074-532-9
eISBN 978-1-78074-533-6

Printed and bound in Great Britain by TJ International Ltd, Padstow, Cornwall

Oneworld Publications
10 Bloomsbury Street
London WC1B 3SR
England

To Jordan and Ella

No continuous thing is divisible into things without parts.

—ARISTOTLE

CONTENTS

INFINITESIMAL

Introduction

A COURTIER ABROAD

In the winter of 1663 the French courtier Samuel Sorbière was presented at a meeting of the newly founded scientific academy, the Royal Society of London. Sorbière, explained Henry Oldenburg, the Society's distinguished secretary, was a friend from the dark days of the civil war, when the king was driven out of England and made his court in Paris. Now, three years after Charles II had been restored to his throne in London, Oldenburg was proud to host his old friend in his true home, and to share with him the exciting new investigations taking place under the Royal Society's roof. For the next three months, Sorbière traveled the land, meeting with political leaders and leading intellectual lights, and even the king himself. Throughout this time, the gregarious Frenchman made the Royal Society his home, attending its meetings and socializing with its fellows. They, for their part, treated him with the greatest respect, and bestowed on him their highest honor: they made him a fellow of the Royal Society.

Whether Sorbière was worthy of this honor is debatable. Although he was a noted physician in his day, and something of a man of letters, even he did not consider himself an original thinker. By his own testimony he was a "trumpeter" rather than a "soldier" in the "warfare of letters," one who did not promote his own ideas but who advertised the ingenious inventions of other men through his far-flung web of acquaintances and correspondents. And it was, to be sure, an impressive network, including some of the greatest luminaries in France, and philosophers and scientists in Italy, the Dutch Republic, and England. Sorbière was of a type familiar in intellectual circles from that day to ours, the man whom everybody knows, though not necessarily respects all that much. Of greater concern to his hosts, however, was the fact that Sorbière was a close friend and the French translator of Thomas Hobbes, a man most members of the Society considered a dangerous subversive, and a threat to religion and the state.

If the powers that be at the Royal Society were willing to overlook these missteps and invite him into their circle, the reason was simple: Sorbière was a man on the rise. In 1650, after years of living in exile in Holland, he'd returned to France and, four years later, abandoned his Protestant faith and converted to Catholicism. At a time when the position of Protestants in France was becoming increasingly precarious, this was a wise choice. Sorbière became a protégé of Cardinal Mazarin, Louis XIV's chief minister, and was admitted to the king's inner circle. He was granted a pension and the title of royal historiographer, and tried to use his influence as a high-ranking courtier to establish a scientific academy in France. His journey to England was meant, in part, as a study of the Royal Society, to determine whether it could serve as a useful model for a similar institution back home. To the grandees of the fledgling Royal Society, always on the lookout for patrons and benefactors, Sorbière was an emissary from the brilliant court of Louis XIV, and hence a man to be treated with the utmost consideration.

If Oldenburg and his fellows hoped to be repaid in kind for the honor they had granted Sorbière, they were quickly disappointed. Mere months after returning home, Sorbière published an account of his experiences in England that showed little appreciation for the country he had so recently visited, stunning his former hosts. In Sorbière's eyes, England suffered from an excess of religious freedoms and an excess of "Republican spirit," both of which undermined established religion and royal authority. The official Church of England, Sorbière wrote, was likely the best of the plethora of sects, because its "Hierarchy inspires People with Respect to those who are Supream over them, and is a support to the Monarchy." But the others—Presbyterians, Independents, Quakers, Socinians,

Mennonites, etc.—are the "pernicious" fruit of excessive toleration and have no place in a peaceful realm.

Sorbière, to be fair, did lavish praise on the Royal Society, and spoke with admiration of the experiments conducted in its halls and of the civility of the debates among its members. He even predicted that "if the advance-project of the Royal Society be not some way or other blasted," then "we shall find a world of people fall into Admiration of so excellent a learned Body." The details of Sorbière's account, however, were far less flattering. He claimed that the Society was divided between the followers of two French philosophers, Descartes and Gassendi, an assertion that offended the English on both patriotic and principled grounds; the Royal Society prided itself on following nature alone, eschewing any systematic philosophy. Sorbière insulted the Society's patron the Earl of Clarendon, Charles II's Lord High Chancellor, by writing that he understood the formalities of the law, but little else, and had "no knowledge of literature." Of the Oxford mathematician John Wallis (1616–1703), one of the Society's founders and leading lights, Sorbière wrote that his appearance inclined one to laughter and that he suffered from bad breath that was "noxious in conversation." Wallis's only hope, according to Sorbière, was to be purified by the "Air of the court in London."

For the Society's nemesis Thomas Hobbes, however, who was also Wallis's personal enemy, Sorbière had only praise. Hobbes, he wrote, was a courtly and "gallant" man, and a friend of "crowned heads," despite his upbringing as a Protestant. Furthermore, Sorbière claimed, Hobbes was the true heir of the illustrious Sir Francis Bacon, the late Lord Chancellor of England and prophet of the new science. This last was the most egregious of Sorbière's offenses in the eyes of the Royal Society grandees. Bacon was venerated in the Society as its guiding spirit and, effectively, patron saint. To have his mantle bestowed on Hobbes was intolerable. As Thomas Sprat, the Society's historian, wrote in a thorough rebuttal of Sorbière's account, there was no more likeness between Hobbes and Bacon than there was "between St. George and the Waggoner."

Sorbière, as it turned out, paid dearly for his perceived ingratitude to his English hosts. He may have cared little for Sprat's insults, hurled from faraway London, but he could not ignore the unpleasant ramifications in the royal court in Paris. France at the time sided with England in its war against the Dutch Republic, and Louis XIV was not pleased that one of his own courtiers was causing diplomatic friction with a useful ally. He promptly stripped Sorbière of his status as royal historiographer and banished him from court. Though the banishment was lifted several months later, things were never the same for Sorbière. He repeatedly tried to ingratiate himself with the king, and when that failed, he traveled to Rome to seek the Pope's patronage instead. He died in 1670, never having regained the status and prestige he had enjoyed on the eve of his journey to England.

Though disastrously ill-timed as far as his career was concerned, Sorbière's *A Voyage to England* expresses views that in many ways were what one might expect from a man in his position. He was, after all, a courtier to Louis XIV, the king most responsible for establishing royal absolutism in France, and whose philosophy of governance was well encapsulated in his (possibly apocryphal) saying "L'état c'est moi." In the 1660s Louis was rapidly concentrating state power in royal hands, and was well on his way to creating a single-faith state, a process completed with the expulsion of the Protestant Huguenots in 1685. If the French court's ambition was to create a nation of "one king, one law, one faith" ("un roi, une loi, une foi"), then certainly Sorbière saw little evidence of that in England. Not only had the English effectively suppressed the true Catholic faith, but they had not even succeeded in replacing it with a single religion of their own. A plethora of sects was undermining the established state religion, and thereby the authority of the king. Personages whose actions during the civil war had suggested dangerous republican tendencies now occupied respectable positions in both church and state, whereas Hobbes, a steadfast royalist whose philosophy supported "crowned heads," was marginalized.

Nor were things better when it came to the personal manners of Englishmen. In France, membership in court society was the highest social as well as political aspiration of any man or woman eager to leave his or her mark. The members of this exclusive society were distinguished by their fashionable dress and refined manners, all designed to set them apart from outsiders and establish their social superiority. Sorbière's English hosts, however, showed little inclination to follow the French example. While some of them—including Royal Society president Lord Brouncker and the noble-born Robert Boyle—were members of the high aristocracy, whose upbringing equaled that of any French courtier, others were not. And as was clear in the case of Wallis, a lack of courtly grace did not disqualify one from a place of honor in the highest intellectual circles. Hobbes, in contrast, had spent a lifetime as a member of aristocratic houses, had adopted their manners, and was therefore a man after Sorbière's own heart. By ridiculing Wallis and praising Hobbes, Sorbière was doing more than merely expressing his personal sensibilities; he was criticizing the lack of courtly refinement in English society and lamenting the fact that, in England, the court did not set the cultural tone of the land as it did in France. When the low mixed with the high, and boors such as Wallis were allowed in high society, what hope was there for court and king to establish their authority? Such mixing would never have been allowed at the court of the Sun King, and only confirmed Sorbière's opinion that a dangerous "Spirit of Republicanism" was lurking beneath the surface of English society.

Hobbes, in Sorbière's view, was everything a cultured man should be: courtly in his manners, a friend and companion to the great men of the realm, a steadfast loyal subject, and a philosopher whose teachings (in Sorbière's opinion) supported the rule of kings. Wallis was quite the opposite: uncourtly and uncouth, and a former Parliamentarian who had made war on his king and who had undeservedly been granted a place of honor by the restored monarch. Little wonder that in the long feud between Wallis and Hobbes, the French monarchist sided with Hobbes. But in his account of their dispute, Sorbière does not dwell on the two men's political or religious differences, but concentrates on something else entirely: "The Argument," he explains, "was about the indivisible Line of the Mathematicians, which is a mere Chimera, of which we can have no idea." For Sorbière, it all boiled down to this: Wallis accepted the concept of mathematical indivisibles; Hobbes (and Sorbière with him) did not. Therein lay the difference.

The notion that a political essayist reviewing the institutions of a foreign land would focus on an obscure mathematical concept seems not only surprising to us today, but outright bizarre. The concepts of higher mathematics appear to us so abstract and universal that they cannot be relevant to cultural or political life. They are the domain of highly trained specialists, and do not even register with modern-day cultural critics, not to mention politicians. But this was not the case in the early modern world, for Sorbière was far from the only nonmathematician to be concerned about the infinitely small. In fact, in Sorbière's day, European thinkers and intellectuals of widely divergent religious and political affiliations campaigned tirelessly to stamp out the doctrine of indivisibles and to eliminate it from philosophical and scientific consideration. In the very years that Hobbes was fighting Wallis over the indivisible line in England, the Society of Jesus was leading its own campaign against the infinitely small in Catholic lands. In France, Hobbes's acquaintance René Descartes, who had initially shown considerable interest in infinitesimals, changed his mind and ultimately banned the concept from his all-encompassing philosophy. Even as late as the 1730s the High Church Anglican bishop George Berkeley mocked mathematicians for their use of infinitesimals, calling these mathematical objects "the ghosts of departed quantities." Lined up against these naysayers were some of the most prominent mathematicians and philosophers of that era, who championed the use of the infinitely small. These included, in addition to Wallis: Galileo and his followers, Bernard Le Bovier de Fontenelle, and Isaac Newton.

Why did the best minds of the early modern world fight so fiercely over the infinitely small? The reason was that much more was at stake than an obscure mathematical concept: The fight was over the face of the modern world. Two camps confronted each other over the infinitesimal. On the one side were ranged the forces of hierarchy and order—Jesuits, Hobbesians, French royal courtiers, and High Church Anglicans. They believed in a unified and fixed order in the world, both natural and human, and were fiercely opposed to infinitesimals. On the other side were comparative "liberalizers" such as Galileo, Wallis, and the Newtonians. They believed in a more pluralistic and flexible order, one that might accommodate a range of views and diverse centers of power, and championed infinitesimals and their use in mathematics. The lines were drawn, and a victory for one side or the other would leave its imprint on the world for centuries to come.

THE TROUBLE WITH INFINITESIMALS

To understand why the struggle over indivisibles became so critical, we need to take a close look at the concept itself, which appears deceptively simple but is in fact deeply problematic. In its simplest form the doctrine states that every line is composed of a string of points, or "indivisibles," which are the line's building blocks, and which cannot themselves be divided. This seems intuitively plausible, but it also leaves much unanswered. For instance, if a line is composed of indivisibles, how many and how big are they? One possibility is that there is a very large number of such points in a line, say a billion billion indivisibles. In that case, the size of each indivisible is a billion-billionth of the original line, which is indeed a very small magnitude. The problem is that any positive magnitude, even a very small one, can always be divided. We could, for instance, divide the original line into two equal parts, then divide each of them into a billion billion parts, which would result in segments that were half the size of our original "indivisibles." That means that our supposed indivisibles are, in fact, divisible after all, and our initial supposition that they are the irreducible atoms of the continuous line is false.

The other possibility is that there is not a "very large number" of indivisibles in a line, but actually an infinite number of them. But if each of these indivisibles has a positive magnitude, then an infinite number of them arranged side by side would be infinite in length, which goes against our assumption that the original line is finite. So we must conclude that the indivisibles have no positive magnitude, or, in other words, that their size is zero. Unfortunately, as we know, $0+0=0$, which means that no matter how many indivisibles of size zero we add up, the combined magnitude will still be zero and will never add up to the length of the original line. So, once again, our supposition that the continuous line is composed of indivisibles leads to a contradiction.

The ancient Greeks were well aware of these problems, and the philosopher Zeno the Eleatic (fifth century BCE) codified them in a series of paradoxes with colorful names. "Achilles and the Tortoise," for example, demonstrates that swift Achilles will never catch up with the tortoise, be the latter ever so slow, if Achilles first has to pass one-half of the distance between them, then one-quarter, one-eighth, and so on. Yet we know from experience that Achilles will catch up with his slower rival, leading to a paradox. Zeno's "Arrow" paradox asserts that an object that fills a space equal to itself is at rest. This, however, is true of an arrow at every instant of its flight, which leads to the paradoxical conclusion that the arrow does not move. Though seemingly simple, Zeno's mind-benders prove extremely difficult to resolve, based as they are on the inherent contradictions posed by indivisibles.

But the problems do not end there, for the doctrine of indivisibles also runs up against the fact that some magnitudes are incommensurable with others. Consider, for example, two lines with lengths given as 3 and 5. Obviously the length 1 is included three full times in the shorter line, and five full times in the longer. Because it is included a whole number of times in each, we call the length 1 a common measure of the line with length 3 and the line with length 5. Similarly, consider lines with lengths of 3½ and 4½. Here the common measure is ½, which is included 7 times in 3½ and 9 times in 4½. But things break down if you consider the side of a square and its diagonal. In modern terms we would say that the ratio between the two lines is $\sqrt{2}$, which is an irrational number. The ancients put it differently, effectively proving that the two lines have no common measure, or are "incommensurable." This means that no matter how many times you divide each of the lines, or how thinly you slice them, you will never arrive at a magnitude that is their common measure. Why are incommensurables a problem for indivisibles? Because if lines were composed of indivisibles, then the magnitude of these mathematical atoms would be a common measure for any two lines. But if two lines are incommensurable, then there is no common component that they both share, and hence there are no mathematical atoms, no indivisibles.

The discovery of these ancient conundrums by Zeno the Eleatic and the followers of Pythagoras in the sixth and fifth centuries BCE changed the course of ancient mathematics. From then on, classical mathematicians turned away from the unsettling considerations of the infinitely small and focused instead on the clear, systematic deductions of geometry. Plato (ca. 428–348 BCE) led the way, making geometry the model for correct rational reasoning in his system, and (according to tradition) carving the words "let no one ignorant of geometry enter here" above the entrance to his Academy. His student Aristotle (ca. 384–322 BCE) differed from his master on many issues, but he too agreed that infinitesimals must be avoided. In a detailed and authoritative discussion of the paradoxes of the continuum in book 6 of his *Physics*, he concluded that the concept of infinitesimals was erroneous, and that continuous magnitudes can be divided ad infinitum.

The turn away from infinitesimals would likely have been final had it not been for the remarkable work of the greatest of all ancient mathematicians, Archimedes of Syracuse (ca. 287–212 BCE). Fully aware of the mathematical risks he was taking, Archimedes nevertheless chose to ignore, at least provisionally, the paradoxes of the infinitely small, thereby showing just how powerful a mathematical tool the concept could be. To calculate the volumes enclosed in circles, cylinders, or spheres, he sliced them up into an infinite number of parallel surfaces and then added up their surface areas to arrive at the correct result. By assuming, for the sake of argument, that continuous magnitudes are, in fact, composed of indivisibles, Archimedes was able to reach results that were well nigh impossible in any other way.

Archimedes was careful not to rely too much on his novel and problematic method. After arriving at his results by means of infinitesimals, he went back and proved every one of them by conventional geometrical means, avoiding any use of the infinitely small. Even so, despite his caution, and his fame as a great sage of the ancient world, Archimedes had no mathematical successors. Future generations of mathematicians steered clear of his novel approach, relying instead on the tried-and-true methods of geometry and its irrefutable truths. For over a millennium and a half, Archimedes's work on infinitesimals remained an anomaly, a glimpse of a road not taken.

It was not until the 1500s that a new generation of mathematicians picked up the cause of the infinitely small. Simon Stevin in Flanders, Thomas Harriot in England, Galileo Galilei and Bonaventura Cavalieri in Italy, and others rediscovered Archimedes's experiments with infinitesimals and began once more to examine their possibilities. Like Archimedes, they calculated the areas and volumes enclosed in geometrical figures, then went beyond the ancient master by calculating the speed of bodies in motion and the slopes of curves. Whereas Archimedes was careful to say that his results were only provisional until proven through traditional geometrical means, the new mathematicians were less timid. Defying the well-known paradoxes, they openly treated the continuum as made up of indivisibles and proceeded from there. Their boldness paid off, as the "method of indivisibles" revolutionized the practice of early modern mathematics, making possible calculations of areas, volumes, and slopes that were previously unattainable. A staid field, largely unchanged for centuries, was turned into a dynamic one that was constantly expanding and acquiring new and unprecedented results. Later on, in the late seventeenth century, the method was formalized at the hands of Newton and Leibniz, and became the reliable algorithm that today we call the "calculus," a precise and elegant mathematical system that can be applied to an unlimited range of problems. In this form, the method of indivisibles, founded on the paradoxical doctrine of the infinitely small, became the foundation of all modern mathematics.

THE LOST DREAM

Yet, useful as it was, and successful as it was, the concept of the infinitely small was challenged at every turn. The Jesuits opposed it; Hobbes and his admirers opposed it; Anglican churchmen opposed it, as did many others. What was it, then, about the infinitely small that inspired such fierce opposition from so many different quarters? The answer is that the infinitely small was a simple idea that punctured a great and beautiful dream: that the world is a perfectly rational place, governed by strict mathematical rules. In such a world, all things, natural and human, have their given and unchanging place in the grand universal order. Everything from a grain of sand to the stars in the sky, from the humblest beggar to kings and emperors, is part of a fixed, eternal hierarchy. Any attempt to revise or topple it is a rebellion against the one unalterable order, a senseless disruption that, in any case, is doomed to failure.

But if the paradoxes of Zeno and the problem of incommensurability prove anything, it is that the dream of a perfect fit between mathematics and the physical world is untenable. On the scale of the infinitely small, numbers do not correspond to physical objects, and any attempt to force the fit leads to paradoxes and contradictions. Mathematical reasoning, however rigorous and true on its own terms, cannot tell us how the world actually must be. At the heart of creation, it seems, lies a mystery that eludes the grasp of the most rigorous reasoning, and allows the world to diverge from our best mathematical deductions and go its own way—we know not where.

This was deeply troubling to those who believed in a rationally ordered and eternally unchanging world. In science it meant that any mathematical theory of the world was necessarily partial and provisional, because it could not explain everything in the world, and might always be replaced by a better one. Even more troubling were the social and political implications. If there was no rational and unalterable order in society, what was left to guarantee the social order and prevent it from descending into chaos? To groups invested in the existing hierarchy and social stability, infinitesimals seemed to open the way to sedition, strife, and revolution.

Those, however, who welcomed the introduction of the infinitely small into mathematics held far less rigid views about the order of the natural world and society. If the physical world was not ruled by strict mathematical reasoning, there was no way to tell in advance how it was structured and how it operated. Scientists were therefore required to gather information about the world and experiment with it until they arrived at an explanation that best fit the available data. And just as they did for the natural world, infinitesimals also opened up the human world. The existing social, religious, and political order could no longer be seen as the only possible one, because infinitesimals had shown that no such necessary order existed. Just as the opponents of infinitesimals had feared, the infinitely small led the way to a critical evaluation of existing social institutions and to experimentation with new ones. By demonstrating that reality can never be reduced to strict mathematical reasoning, the infinitely small liberated the social and political order from the need for inflexible hierarchies.

The struggle over the infinitely small in the early modern world took different forms in different places, but nowhere was it waged with more determination, or with higher stakes, than in the two poles of Western Europe: Italy in the south and England in the north. In Italy it was the Jesuits who led the charge against infinitesimals, as part of their efforts to reassert the authority of the Catholic Church in the wake of the disastrous years of the Reformation. The story of this fight, from its faint glimmers in the early history of the Society of Jesus to the climactic struggles with Galileo and his followers, is told in part 1 of this book, "The War against Disorder." In England, too, the struggle over the infinitely small followed in the wake of turbulence and upheaval—the two decades of civil war and revolution in the middle of the seventeenth century during which England was a troubled land without a king. The drawn-out gladiatorial fight over infinitesimals between Thomas Hobbes and John Wallis was a struggle between two competing visions of the future of the English state. The story of this fight—its roots in the terror-filled days of the revolution, its role in the founding of the leading scientific academy in the world, and its effect on the emergence of England as a leading world power—is found in part 2 of this book, "Leviathan and the Infinitesimal."

From north to south, from England to Italy, the fight over the infinitely small raged across western Europe. The lines in the struggle were clearly drawn. On the one side were the advocates of intellectual freedom, scientific progress, and political reform; on the other, the champions of authority, universal and unchanging knowledge, and fixed political hierarchy. The results of the fight were not everywhere the same, but the stakes were always just as high: the face of the modern world, then coming into being. The statement that "the mathematical continuum is composed of distinct indivisibles" is innocent enough to us, but three and a half centuries ago it had the power to shake the foundations of the early modern world. And so it did: the ultimate victory of the infinitely small helped open the way to a new and dynamic science, to religious toleration, and to political freedoms on a scale unknown in human history.

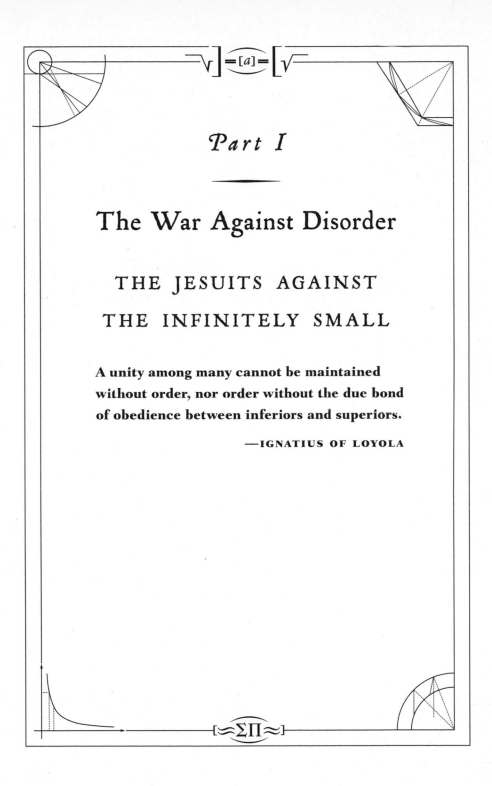

Part I

The War Against Disorder

THE JESUITS AGAINST
THE INFINITELY SMALL

A unity among many cannot be maintained without order, nor order without the due bond of obedience between inferiors and superiors.

—IGNATIUS OF LOYOLA

I

The Children of Ignatius

A MEETING IN ROME

On August 10, 1632, five men in flowing black robes came together in a somber Roman palazzo on the left bank of the Tiber River. Their dress marked them as members of the Society of Jesus, the leading religious order of the day, as did their place of meeting—the Collegio Romano, headquarters of the Jesuits' far-flung empire of learning. The leader of the five was the elderly German father Jacob Bidermann, who had made a name for himself as the producer of elaborate theatrical performances on religious themes. The others are unknown to us, but their names— Rodriguez, Rosco, Alvarado, and (possibly) Fordinus—mark them as Spaniards and Italians, like many of the men who filled the ranks of the Society. In their day these men were nearly as anonymous as they are today, but their high office was not: they were the "Revisors General" of the Society of Jesus, appointed by the general of the order from among the faculty of the Collegio. Their mission: to pass judgment upon the latest scientific and philosophical ideas of the age.

The task was a challenging one. First appointed at the turn of the seventeenth century by General Claudio Acquaviva, the Revisors arrived on the scene just in time to confront the intellectual turmoil that we know as the scientific revolution. It had been over half a century since Nicolaus Copernicus published his treatise proclaiming the novel theory that the Earth revolved around the sun, and the debate on the structure of the heavens had raged ever since. Could it be possible that, contrary to our daily experience, common sense, and established opinion, the Earth was moving? Nor were things simpler in other fields, where new ideas seemed to be cropping up daily—on the structure of matter, on the nature of magnetism, on transforming base metals into gold, on the circulation of the blood. From across the Catholic world, wherever there was a Jesuit school, mission, or residence, a steady stream of questions came flowing to the Revisors General in Rome: Are these new ideas scientifically sound? Can they be squared with what we know of the world, and with the teachings of the great philosophers of antiquity? And most crucially, do they conflict with the sacred doctrines of the Catholic Church? The Revisors took in these questions, considered them in light of the accepted doctrines of the Church and the Society, and pronounced their judgment. Some ideas were found acceptable, but others were rejected, banned, and could no longer be held or taught by any member of the Jesuit order.

In fact, the impact of the Revisors' decisions was far greater. Given the Society's prestige as the intellectual leader of the Catholic world, the views held by Jesuits and the doctrines taught in the Society's institutions carried great weight far beyond the confines of the order. The pronouncements coming from the Society were widely viewed as authoritative, and few Catholic scholars would have dared champion an idea condemned by the Revisors General. As a result, Father Bidermann and his associates could effectively determine the ultimate fate of the novel proposals brought before them. With the stroke of a pen, they could decide which ideas would thrive and be taught in the four corners of the world and which would be consigned to oblivion, forgotten as if they had never been proposed. It was a heavy responsibility, requiring both great learning and sound judgment. Little wonder that only the most experienced and trusted teachers at the Collegio Romano were deemed worthy to serve as Revisors.

But the issue that was brought before the Revisors General that summer day in 1632 appeared far from the great questions that were shaking the intellectual foundations of Europe. While a few short miles away Galileo was being denounced (and would later be condemned) for advocating the motion of the Earth, Father Bidermann and his colleagues were concerning themselves with a technical, even petty question. They had been asked to pronounce on a doctrine, proposed by an unnamed "Professor of Philosophy," on the subject of "the composition of the continuum by indivisibles."

Like all the doctrinal proposals presented to the Revisors, the proposition was cast in the obscure philosophical language of the age. But at its core, it was very simple: any continuous magnitude, it stated, whether a line, a surface, or a length of time, was composed of distinct infinitely small atoms. If the doctrine is true, then what appears to us as a smooth line is in fact made up of a very large number of separate and absolutely indivisible points, ranged together side by side like beads on a string. Similarly, a surface is made up of indivisibly thin lines placed next to each other, a time period is made up of minuscule instants that follow each other in succession, and so on.

This simple notion is far from implausible. In fact, it seems commonsensical, and fits very well with our daily experience of the world: Aren't all objects made up of smaller parts? Is not a piece of wood made of fibers; a cloth, of threads; an hour, of minutes? In much the same way, we might expect that a line will be composed of points; a surface, of lines; and even time itself, of separate instants. Nevertheless, the judgment of the black-robed fathers who met at the Collegio Romano that day was swift and decisive: "We consider this proposition to be not only repugnant to the common doctrine of Aristotle, but that it is by itself improbable, and . . . is disapproved and forbidden in our Society."

So ruled the holy fathers, and in the vast network of Jesuit colleges, their word became law: the doctrine that the continuum is composed of infinitely small atoms was ruled out, and could not be pursued or taught. With this, the holy fathers had every reason to believe, the matter was closed. The doctrine of the infinitely small was now forbidden to all Jesuits, and other intellectual centers would no doubt follow the order's example. Advocates of the banned doctrine would be excluded and marginalized, crushed by the authority and prestige of the Jesuits. Such had been the case with numerous other pronouncements coming out of the Collegio, and Father Bidermann and his colleagues had no reason to think that this time would be any different. As far as they were concerned, the question of the composition of the continuum had been settled.

Looking back from the vantage point of the twenty-first century, one cannot help but be struck, and perhaps a bit startled, by the Jesuit fathers' swift and unequivocal condemnation of "the doctrine of indivisibles." What, after all, is so wrong with the plausible notion that continuous magnitudes, like all smooth objects, are made of tiny atomic particles? And even supposing that the doctrine is in some way incorrect, why would the learned professors of the Collegio Romano go out of their way to condemn it? At a time when the struggle over Copernicus's theory raged most fiercely; when the fate of Galileo, Copernicus's ardent advocate and the most famous scientist in Europe, hung in the balance; when novel theories on the heaven and the earth seemed to pop up regularly, didn't the illustrious Revisors General of the Society of Jesus have greater concerns than whether a line was composed of separate points? To put it bluntly, didn't they have more important things to worry about?

Apparently not. For, strange as it might seem to us, the condemnation of indivisibles in 1632 was not an isolated incident in the chronicles of the Jesuit Revisors, but merely a single volley in an ongoing campaign. In fact, the records of the meetings of the Revisors, which are kept to this day in the Society's archives in the Vatican, reveal that the structure of the continuum was one of the main and most persistent of this body's concerns. The matter had first come up in 1606, just a few years after General Acquaviva created the office, when an early generation of Revisors was asked to weigh in on the question of whether "the continuum is composed of a finite number of indivisibles." The same question, with slight variations, was proposed again two years later, and then again in 1613 and 1615. Each and every time, the Revisors rejected the doctrine unequivocally, declaring it to be "false and erroneous in philosophy . . . which all agree must not be taught."

Yet the problem would not go away. In an effort to keep abreast of the most recent developments in mathematics, teachers from all corners of the Jesuit educational system kept proposing different variations on the doctrine in the hope that one would be tolerated: Perhaps a division into an infinite number of atoms was allowable, even if a finite number was not? Maybe it was permitted to teach the doctrine not as truth but as an unlikely hypothesis? And if fixed indivisibles were banned, what about indivisibles that expanded and contracted as needed? The Revisors rejected all of these. In the summer of 1632, as we have seen, they once again ruled against indivisibles, and Father Bidermann's successors (including Father Rodriguez), when called to pass judgment on it in January 1641, again declared the doctrine "repugnant." In a sign that these decrees had no more lasting effect than their predecessors, the Revisors felt the need to denounce indivisibles again in 1643 and 1649. By 1651 they had had enough: determined to put an end to unauthorized opinions in their ranks, the leaders of the Society produced a permanent list of banned doctrines that could never be taught or advocated by members of the order. Among the forbidden teachings, featured repeatedly in various guises, was the doctrine of indivisibles.

What was it about the indivisibles that was so abhorrent to the Jesuit Revisors in the seventeenth century? The Jesuits, after all, were a religious order—the greatest one of the day—whose purpose was saving souls, not resolving abstract, technical philosophical questions. Why, then, would they bother to proclaim their opinion on so inconsequential a matter, pursue it and its advocates decade after decade, and with the sanction of the highest authorities of the order, make every effort to stamp it out? Clearly the Black Robes, as the Jesuits were popularly known, saw something in this apparently innocuous thesis that is completely invisible to the modern reader—something dangerous, perhaps even subversive, that could threaten an article of faith or core belief the Society held dear. To understand what this was, and why the largest and most powerful religious order in Europe took it upon itself to eradicate the doctrine of indivisibles, we need to go back a century, to the founding days of the order in the early sixteenth century. It was during that time that the seeds of the Jesuit "war on indivisibles" were sown.

THE EMPEROR AND THE MONK

In the year 1521 the young emperor Charles V convened a meeting of the estates of the Holy Roman Empire in the west German city of Worms. Only two years past his election to his high office, Charles was titular head of the Holy Roman Empire, commanding the allegiance of its princes and vast populace. In fact, he was both less and more than that: less because the so-called "empire" was in reality a patchwork of dozens of principalities and cities, each fiercely protective of its independence and as likely to oppose as to aid its imperial lord in time of need; and more because Charles was no ordinary prince; he was a Habsburg, a member of the greatest noble family the West has ever known, with possessions extending from the coast of Castile to the plains of Hungary. Consequently, Charles was not only the elected emperor of Germany, but also, by birthright, the king of Spain and the duke of portions of Austria, Italy, and the Low Countries. Moreover, in those very years, Castile was fast acquiring new territories in the Americas and the Far East, making Charles, in a phrase from the time, "the emperor in whose lands the Sun never sets." And though Francis I of France and Henry VIII of England might have bridled at the suggestion, to his contemporaries as well as to himself, Charles V was the leader of Western Christendom.

In the winter of 1521, however, it was his fractured German empire, not his vast overseas possessions, that were chiefly on the emperor's mind. It had been three and a half years since Martin Luther, an unknown Augustinian monk and professor of theology, nailed a copy of his Ninety-Five Theses to the door of the Castle Church in Wittenberg. The theses themselves were narrowly focused, confronting what Luther saw as an unconscionable abuse practiced by the Church: the sale of "indulgences," which were guarantees of divine grace, absolving the purchasers of their sins and sparing them the torments of purgatory. Luther was far from alone in denouncing the sale of indulgences, which was one among many Church practices that were routinely condemned as abuses by both clerics and laypeople. Nevertheless, Luther's open challenge to Church authorities struck a nerve with both scholars and the common people like nothing before it. Over the following months, with the aid of the newly invented printing press, the theses were disseminated all across the Holy Roman Empire, and were enthusiastically received nearly everywhere.

If this had been where things ended, then the affair would have been of no concern to Charles V. Like many in his day, Charles, too, was distressed by the more egregious practices of the Church, and he may even have felt some sympathy toward the audacious monk. But events soon acquired a momentum of their own. Alarmed by Luther's success, his

Augustinian superiors called him to account at a meeting in Heidelberg, but by the time he left, he had converted many of them to his position. When he was then summoned to Rome, he sheltered under the protection of his prince, the elector Frederick the Wise of Saxony, who arranged for a hearing for him in Germany instead. In an effort to discredit this irksome critic, Church authorities sent the Dominican professor Johann Eck of Ingolstadt, a professional debater and theologian, to confront Luther. The two met for a public debate in 1519, in which Eck skillfully maneuvered his opponent into admitting to clear heresies: that divine grace is granted to believers through faith alone, not through the sacraments of the Church; that the Church is a purely human construct and holds no special power to mediate between men and God; and that its supreme head, the Pope, is fundamentally an impostor. Luther made no apologies for his beliefs; Eck denounced him as a heretic.

Unfortunately for Church leaders, this designation did nothing to slow down the zealous Luther. In 1520 he published three treatises that outlined his basic doctrines in deliberate defiance of established teachings. No longer a critic, he was now a rebel, openly calling for the overthrow of the Church hierarchy and institutions. His influence continued to spread, first in Wittenberg, then in Saxony, and soon clear across Germany and beyond. Everywhere, it seemed, Luther was acquiring followers in all classes and stations of life—men and women, nobles and peasants, country people and city dwellers—all of whom saw him as a leader of a religious awakening that would displace the ossified and corrupt Church of Rome. At long last growing alarmed at the fast-deteriorating situation, Pope Leo X excommunicated Luther, but by this time the drastic action had little effect. Luther's teachings were spreading like wildfire throughout German lands.

It was at this time, spurred by the threat of religious schism, that Charles V entered the fray. Two centuries later the French philosophe Voltaire would mock the empire as "neither Holy, nor Roman, nor an Empire," but to Charles, his realm was holy indeed. As the secular leader of Christendom, and a devout Christian himself, he saw it as his sacred duty to preserve the Church and the spiritual unity of his people. Though Holy Roman emperors have for centuries vied with popes for supremacy in Europe, their squabbles on occasion resolved only by open warfare, it was clear to Charles that the former could not do without the latter.

After all, it was the Pope who, since the days of Charlemagne, crowned emperors, and the Church that gave legitimacy and purpose to the office of emperor. The notion of an empire without a Roman Church, or an emperor without a pope, was unthinkable to Charles. To put his domains in order, and stop the spread of the Lutheran heresy once and for all, he called for a "diet"—a meeting of the estates of the empire.

When the diet convened in the city of Worms in January of 1521, Charles sent Luther a summons to appear before the emperor and the estates and account for his actions. Despite Charles's guarantee of his safety, many of Luther's friends cautioned him against placing himself in his enemies' power and advised him not to go. Nevertheless, in April, Luther arrived in the city and was promptly called before the assembled notables. There he was immediately presented with a list of his heretical doctrines and asked to acknowledge them and recant. Luther was surprised; he had expected to be allowed to argue his case, and was unprepared for the swiftness of the attack. He managed only to ask for a day's reprieve to consider the matter, and Charles, the chivalrous Christian emperor, granted the request. But the next day, Luther was ready. He willingly acknowledged his beliefs, even in the face of hostile questioning and fierce denunciations. When pressed to recant, he answered calmly, "Here I stand; I can do no other; God help me; amen."

With these words Luther ensured the failure of Charles's mission to stamp out heresy in his German lands, but he also did much more: he sealed the fate of Western Christendom. For over a thousand years the Roman Church had reigned supreme in western Europe. It had witnessed the rise and fall of empires, invasion and occupation by infidels, heresies large and small, plague and pestilence, and ruinous wars of king against king and emperor against pope. Through it all, the Church had survived, thrived, and expanded its reach until, by the sixteenth century, its dominion stretched from Sicily to Scandinavia and from Poland to Portugal, and to beachheads in the New World. From baptism to last rites, the Roman Church oversaw the lives of Europeans, giving order, meaning, and purpose to their existence, and ruling on everything from the date of Easter to the motion of the Earth and the structure of the heavens. To the people of western Europe, regardless of nation, language, or political allegiance, the fabric of life itself was inextricably bound to the Roman Church.

But when Luther took his stand at the Diet of Worms, this spiritual and cultural unity came to an abrupt end. By proudly proclaiming his heretical beliefs, he renounced the authority of the Roman Church and led his followers along a new religious path. By openly defying both Pope and emperor before a public gathering of the great men of the empire, he burned his bridges, and eliminated any chance of reconciliation. What up to that point could have been viewed as an internal rebellion within the Church now became a schism in which two rival faiths confronted each other in open hostility. On the one side stood the followers of the old Church; the Pope; and his secular sword, the emperor. On the other, the adherents of the new "Protestant" Church, which claimed direct descent from the ancient apostolic church and rejected the Roman faith as a monstrous aberration. The spiritual unity of the West was shattered with one blow, and any realistic hope of healing the rift through concili-ation or threats was at an end. Luther and his followers refused to concede their errors or surrender to the might of the empire. Conse-quently, they had to be subdued by force of arms.

DECLINE INTO CHAOS

For the next thirty-four years of his reign, Charles V tried to accomplish precisely that. Though all too often distracted by threats from his Euro-pean rivals and the Ottoman sultan, he nevertheless carried on a consis-tent campaign to suppress the cancer of Protestantism that was spreading through his lands. But it was too late. Not only was the new faith gaining adherents by the day among the populace, but great princes of the em-pire were also rallying to Luther, and establishing his church in their territories. First were the electors of Saxony, Frederick the Wise and his successors, who had been Luther's protectors from the beginning. Next was Albrecht of Hohenzollern, grand master of the Teutonic knights, who refashioned himself as the first duke of Prussia and laid the foundations of what would become the greatest Protestant power in Germany. The elec-tor Philip of Hesse followed suit, as did the margrave of Brandenburg, the dukes of Schleswig and Brunswick, and many smaller potentates of the empire. The great imperial cities (Nuremberg, Strassburg, Augsburg) also sided with Luther, broke with the Pope, and established their own reformed churches. By the mid 1520s it seemed nothing could resist the rising tide of Lutheranism.

If the rupture of the empire wasn't bad enough, it soon became clear the dissolution of Christendom would not stop there. In the early 1520s a cleric at Zurich Cathedral named Huldrych Zwingli began preaching fiery sermons denouncing the wickedness of Rome and advocating doctrines even more radical than Luther's. Within a few years he rallied Zurich, and then the neighboring Swiss cities of Bern and Basel, to his cause. Zwingli's death in a battle against the Catholic forest cantons of the Swiss Confederacy in 1531 brought a temporary halt to the spread of his radical vision, but by the late 1530s a new beacon of reform had emerged in Geneva. In 1536, John Calvin launched his long campaign to make Geneva into a shining example of the purest Protestant faith, and of upright public and personal morality. Over the following twenty years, Calvin managed to transform Geneva into a strict theocracy in which no individual action was beyond the scope of religious oversight or censure. And though the example of Geneva might appear unattractive to us, reminiscent as it is of some of the darker theocratic regimes of our own day, contemporaries judged it differently. Calvin's city was glorified as the "city on the hill," a shining example of what could be achieved through religious fervor, moral rectitude, and hard work. Aspiring reformers from across Europe flocked to the city to learn from Calvin how it was done, and to spread his teachings to their native lands. Thanks to the example of Geneva, Calvin's brand of Protestantism, set down in his *Institutes of the Christian Religion*, became the most dynamic and influential movement of the Reformation from the 1540s onward. Even without the princely support that had institutionalized Luther's reforms, Calvin captured millions of converts, from France and England in the west to Poland and Hungary in the east.

Meanwhile, disasters kept piling up for the Roman Church, as not only cities and territories but entire kingdoms were lost to Protestantism. In 1527 the Swedish king Gustavus Vasa adopted Lutheranism and, over the following years, established it as a national church. Less than a decade later, Frederick I, a north German prince who had become king of Denmark, drove out the bishops, abolished the monasteries, and installed Lutheranism as the state religion. Since Norway was at the time under Danish suzerainty, and Finland was a province of Sweden, this made all Scandinavia into the Protestant stronghold it remains to this day.

In England, welcoming the Reformation was initially more of a pragmatic than a spiritual choice. Henry VIII had stood faithfully by the Roman Curia in the early years of the Reformation and even authored an anti-Lutheran treatise that earned him the title Defender of the Faith from Pope Leo X. But as the years went by, Henry grew restless as his wife, the Spanish princess Catherine of Aragon, failed to provide him with a male heir. Resolved to replace Catherine with her charismatic lady-in-waiting Anne Boleyn, he appealed to Pope Clement VII to annul his marriage. Clement, eager to maintain close relations with his royal champions, would likely have granted Henry's request, had it not been for the fact that the queen was Charles V's maternal aunt. Charles made it clear that any attempt to cast Catherine aside would be a personal affront to his honor, and Pope Clement VII could not defy his chief protector. He denied the petition, prompting Henry to sever ties with Rome, marry Anne, and declare himself the head of an independent "Church of England" in 1534.

Henry had no interest in the teachings of the continental reformers, and he intended only to replace the authority of the Pope with his own. Nevertheless, once the English church broke with Rome, the trend toward Protestantism proved irreversible. Under Henry's son, the boy king Edward VI (1547–53), the English reformation veered toward radical Protestantism, only to reverse course under Edward's half-sister (and Catherine's daughter) Mary I (1553–58), who restored Catholicism in her tumultuous five-year reign. It was only when Anne's daughter, Elizabeth I (1558–1603), ascended the throne that Protestantism was established as the state religion once and for all. Under the Thirty-Nine Articles of 1563, the Church of England retained many of the outward forms of the Roman Church that were favored by Henry, including the office of bishop, the sacraments, and the worship in grand and lavishly decorated churches and cathedrals. Doctrinally, however, the Church of England looked not to Rome but to Geneva, adopting the core teachings of John Calvin. To the Holy See, England was irretrievably lost.

As the Reformation spread, it soon became clear that religious truth was far from the only thing at stake. With the Pope denounced, the emperor ignored, and all the established authorities questioned and ridiculed, the entire social order came under scrutiny, and the threat of revolution hung in the air. Respectable reformers such as Luther and Calvin, and the conservative kings and princes who backed them, struggled mightily to contain the revolutionary passions set loose by the Reformation, but not always successfully. As early as 1524 the peasants of southern Germany rose in revolt against their princes, demanding greater freedoms and a greater say in the rule of the land. They declared themselves followers of Luther, believing that his overthrow of the spiritual authority of the Roman Church was but a prelude to the overthrow of the social and political order it supported. The socially conservative Luther, however, was horrified at what he saw as a profound misunderstanding and misuse of his doctrines and fiercely denounced the uprising in a tract, *Against the Murderous, Thieving Hordes of the Peasants*. Though the uprising was crushed within the year by the combined forces of Catholic and Protestant princes, the fear that religious reformation might spell social revolution had already taken root.

The dread of social upheaval continued to haunt the Reformation as more and more reformers and would-be prophets openly questioned the established truths and challenged the authority of the powers that be. Many were peaceful, such as the reformer of Strasbourg Martin Bucer, or the saintly wanderers Caspar von Schwenckfeld and Sébastian Franck. But others were not. Thomas Müntzer was an early follower of Luther, until he broke with him over Luther's embrace of the princes' power and the existing social order. In 1524, Müntzer joined the peasants' uprising, preaching to his followers that the end of days was at hand and calling for the blood of princes. He was captured in 1525, tortured, and killed, but his legacy was still at work ten years later, when a group of radical Anabaptists took control of the city of Münster in northwestern Germany. Unlike the mainstream reformers, whose churches included all members of a community, the Anabaptists insisted that only they were the elect, the true church of God, to the exclusion of all others. At Münster they showed just how dangerous such a doctrine can be when it gains possession of earthly power. Under the leadership of Jan Bockelson of Leyden, the Anabaptists imposed a reign of terror in the city, killing or driving out anyone who stood in their way. When Münster's former Catholic bishop, backed by the Lutheran elector of Hesse, laid siege to the city, Bockelson declared himself the Messiah, abolished private property, and instituted polygamy. In 1535 the forces of the bishop and the prince finally overcame the fierce resistance of Bockelson's fanatical followers, and exacted a bloody revenge on the Anabaptists and anyone remotely suspected of association with them. But across Europe, fear of an impending collapse of all social hierarchy and order only deepened.

To many Europeans in those years, it seemed as if the demons of hell had risen from the underworld to spread misery and confusion upon the land. The old Church that had provided meaning, solace, and certainty to its members since time immemorial was being torn apart by an ever-increasing number of competing creeds. Every day seemed to bring more news of lands in the grip of religious turmoil, with every truth challenged and every certainty gone. The rupture of the Church was followed by political division, as Catholic and Protestant princes faced each other across the religious divide. And beneath the religious and political division lurked the nightmare of a social revolution that would sweep away the entire social order, the only one the people of that time had ever known. It was a time of strife and chaos, and to most Europeans one of debilitating confusion and uncertainty: With all old certainties challenged or discredited, and new ones announcing themselves by the day, how was one to know the difference between Truth and Error? Between the path to heaven and the path to hell?

The institution that, in the eyes of most Europeans, was charged with providing the answers and resolution to these questions was, inevitably, the Papacy in Rome. As the Pope was vicar of Christ on earth and spiritual leader of Western Christendom, it was his lands and his people that were either lost in confusion or gripped by the alien certainties of sectarians and schismatics. It was therefore the duty of the Pope to step into the breach, arrest the progress of the Protestant heresy, and return unity, order, and certainty to Christendom. Sadly for the Roman Church, however—catastrophically, even—the men who occupied the seat of St. Peter during those years were signally ill equipped to deal with the crisis that confronted them.

In many respects, the popes of the early sixteenth century were impressive men. Scions of leading Italian families, they were intelligent and highly cultured, and earned a place in history as the greatest patrons of Renaissance art. Popes Julius II (1503–13), Leo X (1513–21),

Clement VII (1523–34), and Paul III (1534–49) commissioned paint-
ings, frescoes, and sculptures by Michelangelo, Rafael, and Titian;
churches and palaces from architects Sangallo and Bramante. They are
responsible for some of the greatest works in the Western tradition, such
as St. Peter's Basilica and Square and the ceiling of the Sistine Chapel.
But faced with the greatest crisis in the history of the Church, they
found themselves helpless. Competent administrators though they were,
they possessed neither the broad vision nor the spiritual authority required to confront the challenge of Protestantism.

The problem was that the Renaissance popes were not, first and
foremost, leaders of Christendom, but rather Italian princelings, whose
loyalties were primarily to their families and clans. Julius II belonged to
the powerful della Rovere clan of Rome, Leo X and Clement VII were
both members of Florence's ruling Medici family, and Paul III a scion of
the ancient Tuscan Farnese family, soon to become the dukes of Parma.
For each of these clans, having one of their own elevated to Pope was
not only a tremendous honor but also an opportunity to amass wealth
and power that might never be repeated. Popes were expected to take
care of their own, to lavish their relatives with territories, titles (both
secular and ecclesiastical), gifts, and incomes. Fully aware that they
would never have obtained their high station without the sponsorship of
their families, the popes readily obliged, making the reign of each Pope
a race against time to accumulate as many possessions and titles for his
family as possible. It was a sorry spectacle of nepotism and greed reminiscent of some of the most corrupt regimes of the developing world
in our own day. It hung like a malodorous cloud around the Holy See,
and undermined any effort by the Pope to exercise spiritual and moral
authority.

Moreover, in addition to being nominal heads of Christendom and
patriarchs of greedy and acquisitive families, Renaissance popes were
also the rulers of a substantial territorial state in central Italy. In their
efforts to consolidate and expand their possessions, the popes became
key players in the cutthroat politics of the Italian peninsula, making
use of all the means at their disposal—from diplomacy to warfare to
outright treachery—to advance their interests. So notorious were they
for their amorality and ruthlessness in Italian politics that it was Cesare
Borgia, nephew to Pope Alexander VI (1492–1503) and Alexander's chief

military commander, who served as Machiavelli's model of a cunning and brutal prince.

The popes' active engagement in Italian power struggles not only undercut their spiritual standing, but also hamstrung them politically. In their efforts to protect their domains, the popes had to contend with the rising power of the national states of France and Spain, both of which sought to dominate the Italian peninsula, and each of whom possessed military might and resources on a scale that could never be matched by an Italian prince. The only hope of maintaining the independence of the Papal States was to play the two kingdoms against each other, never allowing either one to gain a permanent victory. The popes managed this delicate dance quite successfully for several decades, albeit at the expense of the people of Italy, who suffered repeated invasions and counterinvasions by their mightier neighbors. But disaster finally struck in 1527, amid one of the periodic wars between Charles V, in his capacity as king of Spain, and Francis I of France. Charles's troops, who had not been paid in months, mutinied, and sacked the city of Rome; the murder, rape, and looting went on for weeks. Pope Clement VII escaped the Vatican just in time, and holed up in the nearby fortress of Castel Sant'Angelo as the carnage swirled around him. He ultimately surrendered to the emperor, paid a ransom for his own life, and conceded extensive territories to Spain. The humiliated and much-diminished Pope was to remain effectively a client princeling of the emperor for years to come.

The upshot of all this was that when confronted with the challenge of the Reformation, the Renaissance popes had no answer. Leo X first attempted to use the most tried-and-true weapon in the papal arsenal by excommunicating Martin Luther, but this had little effect. The pleasure-loving Medici prince simply did not possess the moral stature to face down the upright Luther, and his pronouncements carried little weight. The next option for the popes was to rely on the military might of the emperor to bring the schismatics to heel, and Charles was more than willing to take on this role. The popes, however, from Leo X onward, worried that throwing their hat in with the empire meant abandoning the strategy of playing off the Habsburgs against Valois of France. Calling on Charles would effectively end the independence of the Papal States, and reduce the Pope's temporal power to nothing. So, as Charles V struggled for decades to suppress the Protestant heresy and restore unity to Christendom, he did so with either the grudging support of the Holy See or, just as often, its open enmity. To contemporaries, it seemed that the popes would rather see all Christendom torn to shreds than surrender even a sliver of their power in Italy.

In 1540 the fires of the Reformation were still spreading unchecked through the domains of the Roman Church, and lands that had been under the sway of Rome for centuries were falling away one by one. The commonality of faith and ritual that had unified Western Christendom was replaced by a cacophony of competing creeds, each denouncing the others as impostors or worse. As chaos, war, and subversion ruled the land, the Pope proved helpless to put out the fire, but was as intent as ever on amassing titles and incomes for his relatives and protecting his territorial interests. With schism on the ground and corrupt leadership at the top, any objective observer of the European scene in 1540 would likely have concluded that the days of the ancient Church of Rome were numbered.

But on September 27 of that year, at the height of the storm, Pope Paul III took a small administrative step that seemed to bear little relation to the great events of the day: he approved a petition from a group of ten priests to form a religious company dedicated to serving the Pope and the Church. Though hardly noted at the time, it may have been the single most important step taken by the Papacy to save the Roman Church from dissolution. In his bull announcing the new order, Paul also approved the name requested by the group for their new association: they called it the Society of Jesus.

A RAY OF HOPE

The Society of Jesus, or more commonly, the Jesuit Order, was the creation of one man, the Spanish nobleman Ignatius of Loyola. Born in 1491 to an old aristocratic family in the Basque country, Ignatius spent his early years as a gentleman courtier in the entourage of Ferdinand of Aragon. Though reputedly a good Christian, the handsome Inigo, as he was then called, focused his energies on the arts of courtly refinement and romantic love, rather than religious devotion. Heir to the martial tradition of his ancestors, and an ardent reader of the chivalric literature of his day, he aspired more than anything to live out his dreams of military glory. His opportunity finally came in the spring of 1521 in the Spanish city of Pamplona, just a few short weeks after Luther took his stand at Worms in a different corner of Charles V's empire. With French forces advancing on the city and the Spanish army in retreat, Ignatius convinced the local commander to stand his ground and refuse the French demand to surrender. According to Jesuit lore, when the besiegers breached Pamplona's walls, Inigo stood unyielding in their path, but was immediately cut down, and the city was overrun. Close to death, he was treated kindly by the French and delivered to his family's castle of Loyola.

The ten months Ignatius spent convalescing at his family seat can rightly be considered a turning point in the history of Christianity. Starved for entertainment, and with no romances of chivalry within reach, Ignatius began reading the lives of the saints, and was affected to his very core. The saints, he realized, were God's own army in an eternal struggle against the Devil for possession of the human soul. Here was a war truly worth fighting, and Ignatius was determined to join in. As soon as he was physically able, he would make a pilgrimage to Jerusalem and dedicate his life to God's service. As if to confirm him in his new vocation, he was rewarded one night with a mystical vision of the Virgin Mary.

In the winter of 1522, Ignatius walked away from his convalescent bed a changed man. Gone was the elegant courtier who had spent his days in pursuit of women and martial glory. In his place was a holy pilgrim, sworn to undertake any hardship and deprivation that would come his way in spreading the word of God. Before setting off on his journey to Jerusalem, he spent a year in the small town of Manresa, where he meditated, begged for his sustenance, and had visions of God the Father, the Son, and the saints. He also wrote the first draft of *The Spiritual Exercises*, his manual of meditation that would become the cornerstone of the training and formation of Jesuits for centuries to come. When he finally reached the Holy Land, he spent only nineteen days there. The Franciscan friar in charge of the holy sites grew alarmed at this strange pilgrim's zeal, and unceremoniously sent him home.

Thwarted, Ignatius returned to Spain and embarked on a systematic course of study in theology at the great Spanish universities of Barcelona,

Alcalá, and Salamanca. At thirty-two, he was much older than his class-
mates, and not a young man by the standards of the time. He struggled
in his studies, but nevertheless made a deep impression on his fellow
students in his devoutness and self-imposed poverty. He earned a reputation as a mystic and a spiritual counselor, and acquired a band of loyal
followers who had undergone the course of meditations in his *Spiritual
Exercises*. His success brought him to the attention of the Spanish Inquisition, which imprisoned him for a spell while investigating him for suspected heresy. Though ultimately released, Ignatius concluded that he
could not safely resume his work in Spain, and in 1527 he moved to
Paris to continue his studies at the Sorbonne.

It was at the Sorbonne, among his fellow students, that Ignatius found
the men who would form the kernel of the Society of Jesus. Within a few
short years he had surrounded himself with a tight-knit group of Spanish, Portuguese, and French theology students, all much younger than
he, who viewed him as their undisputed leader in all things, spiritual
and worldly. Along with his followers, he once again determined to travel
to Jerusalem, with the goal of preaching Christianity to the Muslims of
the Holy Land. This time, however, with a touch of realism acquired on
his earlier pilgrimage, he included a backup plan: if for some reason it
proved impractical for his group to travel to Jerusalem or remain there,
they would journey to Rome instead and place themselves at the service
of the Pope.

They never made it to Jerusalem. In 1534 they gathered in Venice to
wait for a ship to the Holy Land, but were stranded due to a lack of
funds and the war between Charles V and the Ottoman sultan. As they
waited, Ignatius's band spent their time preaching the word of God and
serving the poor, sick, and dying in Venice and nearby towns. By 1539,
with hopes for the journey fading, they decided to formalize their association by establishing a new religious order, one dedicated to serving the
Church and the Pope in any corner of the world. The Society, as Ignatius announced in his petition to the Pope, would be open to "whoever
desires to serve as a soldier of God beneath the banner of the cross." It
would be the Pope's own army.

THE CHILDREN OF IGNATIUS

It took almost a year, but Paul III did ultimately approve the Society of Jesus. Showing his doubts, he limited the number of the new order to a mere sixty, but the restriction was soon repealed as the order grew and prospered. Indeed, the early growth of the Society of Jesus was nothing less than spectacular. Only ten men, all close intimates, elected Ignatius as the Society's first general in 1540. But by the time of the founder's death in 1556, the ranks of the order had grown a hundredfold, to one thousand. A decade later the Society comprised thirty-five hundred members, and at the death of General Acquaviva in 1615, no fewer than thirteen thousand men had taken Jesuit orders. Thereafter, the Society's growth was still impressive, if somewhat slower, reaching twenty thousand at the turn of the eighteenth century. Through it all, the Society never compromised on the quality of the new recruits in order to expand its numbers. From the beginning, Ignatius had insisted that all candidates be rigorously screened before being accepted as novices in the Society. For those accepted, the road from novice to full membership was long and arduous, lasting years and sometimes even decades. The Jesuits never relaxed these standards, even though no other religious order required anything remotely as demanding. Despite this, or perhaps because of it, the Jesuits never lacked for volunteers of the highest social and intellectual caliber.

Many of the Society's early leaders came from ancient and noble families, as had Ignatius himself and his companion from the Sorbonne Francis Xavier (1506–52). The order's third general, Francis Borgia (1510–72) had been Duke of Gandia in Castile before taking orders (as well as great-grandson of the notorious "Borgia Pope," Alexander VI), and Claudio Acquaviva was the son of the Duke of Arti in the Kingdom of Naples. Other Jesuits came from humbler origins, but distinguished themselves as the outstanding intellectuals of the age. Such, for example, were the Spanish theologians Francisco de Toledo (1532–96) and Francisco Suárez (1548–1617), and the Venetian Robert Bellarmine (1542–1621). Christopher Clavius (1538–1612), Gregory St. Vincent (1584–1667), and André Tacquet (1612–60) were leading mathematicians; Christoph Grienberger (1561–1636) and Christoph Scheiner (1573–1650), prominent astronomers; and Athanasius Kircher (1601–80) and Roger Boscovich (1711–87), trendsetting natural philosophers. And no list of prominent Jesuits can leave out the brilliant Matteo Ricci (1552–1610), who traveled to China to spread the word of God and became a leading scholar and exponent of Western learning at the Ming imperial court. This is only a small sample, but it is enough to justify the assessment of the French philosopher and essayist Michel de Montaigne, who visited the Jesuits in their Roman headquarters in 1581. He called the order "a nursery of great men."

The Jesuits, however, were far more than an association of impressive individuals. They were a highly trained and disciplined collective, honed into a powerful instrument in a single-minded quest: to spread the teachings of the Catholic Church, expand its reach, and bolster its authority. This was the case from the beginning, when Ignatius and his band of followers first offered to serve the Pope in any corner of the world, imagining themselves preaching the word of God to Muslims in the Holy Land. Although that mission never materialized, it was not long before the Jesuits distinguished themselves with outstanding missionary work on four continents. Already in 1541, Francis Xavier set out from Portugal on a mission that would take him to Goa in India, Java, the Moluccas, and Japan, preaching the Gospel and setting up missions wherever he went. He died in 1552 while awaiting transport to China, where he hoped to convert the most populous nation in the world to the Roman faith. Other Jesuits, meanwhile, traveled to Mexico, Peru, and Brazil, where they joined with Dominican and Franciscan friars in their efforts to Christianize the New World. They worked zealously and efficiently, established residences and missions, cared for the souls of the new settlers, and worked tirelessly to convert the native people of the Americas.

Nevertheless, the crucial impact of the Jesuits lay in dealing with heathens much closer to home. For, in the turbulent years of the Reformation, when the very survival of the old Church hung in the balance, the Jesuits became the elite vanguard of Roman Catholicism, dedicated to holding the line against a Protestant tide that appeared to carry all before it. With remarkable skill, dedication, and an energetic, enterprising spirit, they led a stunning Catholic resurgence that not only halted the spread of the Reformation but won back for the Pope many territories that had seemed lost forever. They were just as Ignatius had imagined them: God's own army battling His enemies, leading a movement of Catholic revival that became known as the Counter-Reformation.

It was the vision of their founder that made the Jesuits so formidable an instrument in the service of the Pope. Already in *The Spiritual Exercises* of 1522—nearly two decades before the Society's official formation—Ignatius demonstrated the inner paradox that would shape the Jesuits for centuries. In the first instance, the *Exercises* is a mystical text, intended to elevate readers above their worldly surroundings and bring them to an ecstatic union with God. The history of the medieval Church is rife with charismatic mystics who, like Ignatius, had visions of Christ and the Virgin and who ascended to a higher, and even divine, plane of existence. In their writing, mystics such as Joachim of Fiore and Catherine of Siena attempted to share something of their experience with their followers, and in that regard Ignatius's tract was quite typical.

But the *Exercises* is something else as well: a carefully detailed practical manual on how to achieve union with God. The prescribed course of meditation is divided into four "weeks," though they need not correspond precisely to seven days. Each week's meditations have a different focus, from the nature of sin and the torments of hell in the first week to the sufferings of Christ and the Resurrection in the fourth. The "exercitant" must follow these directions precisely, with an open heart and a will to renounce selfishness and accept God's proffered grace. The road to God, as mapped out in the *Exercises*, is not a single mysterious leap from our fallen world to godly heavens, explainable only through divine grace. It is, rather, a long and arduous journey requiring discipline, dedication, unquestioning trust in the guidance of one's superiors, and strict obedience to their directions.

The tension between ecstatic mysticism and rigorous discipline, the core of the *Exercises*, makes it profoundly different from other mystical texts, which focus on the glory of union with God but offer no road map for how to achieve it. And it is precisely this paradox that animated the Society of Jesus and made it the powerful and effective tool that it was in the hands of the Papacy. For the Jesuits were unequivocally mystics: each novice entering the Society went through the course of *The Spiritual Exercises* and experienced the blissful union with God that is its culmination. Thereafter he would act with the unquestioning confidence that is the province of those who encountered God and knew what He wanted of them. But whereas traditional mystics were led to a life of solitude and inner contemplation, the Jesuits projected their inner confidence onto the world, proceeding with discipline, order, and endurance. The result was that the Jesuits presented a unique combination of traits that made them into one of the most effective organizations, religious and otherwise, in the history of the world: the zeal and certainty of the mystic, and the rigid organization and focused purpose of an elite military unit.

In addition to establishing the order's guiding principles, Ignatius also put in place the mechanisms that would turn those principles into reality. The greatest challenge, he recognized, was to create a body of men who would be unquestioningly committed to the Society and its goals, and willing to dedicate their entire lives to both. Even a brilliant and highly moral individual might be rejected if the selection committee determined that he was overly individualistic and therefore unsuited to life in a disciplined collective. Once admitted, a young man was separated from his former life and underwent a two-year novitiate in which he was inculcated with the Society's ideals of poverty and service. He would practice the complete sequence of *The Spiritual Exercises* and serve in the Society's far-flung missions, colleges, and residences. Above all, he was required to accept without question the authority of his superiors, and to follow their directions in things large and small.

At the end of the two years, the novices took the monastic vows of poverty, chastity, and obedience. For those who were not expected to be ordained as priests, this was the end of their formal training. They would become "approved coadjutors" and, years later, "formed coadjutors," serving perhaps as administrators, cooks, or gardeners, full members of the Society but of a lower grade than their ordained brethren. Novices destined for the priesthood, however, would become "scholastics," undertaking years of advanced studies in Jesuit institutions. Along the way, they would be ordained as priests, and would also take several years off from their studies to teach incoming students. Once they completed their studies, the scholastics would go through another year of "spiritual formation," at the end of which they would pronounce their final vows. Some would once again pronounce the three traditional vows and become "spiritual coadjutors." But those judged the most outstanding in both learning and character would add a fourth vow, unique to the

Jesuits, professing absolute personal obedience to the Pope. These men were known as the "professed," and formed the order's unchallenged elite. Overall, this long process, lasting between eight and fourteen years, produced the kind of individuals Ignatius had envisioned: intelligent, energetic, and disciplined. Together they were a tight-knit brotherhood, bound by deep identification with the goals of the Society, a strong sense of camaraderie, and pride in belonging to an elite corps in the service of Christ and the Church.

The Jesuits, however, were not just a brotherhood of affection and solidarity; they were also a strictly organized top-down hierarchy, built to operate as smoothly and efficiently as a modern military unit. At the apex was the superior general, invariably a professed member, elected for life by the order's general congregation. His powers within the order were unlimited, and he was free to appoint or dismiss any Jesuit from any position within the order. Below him were the provincial superiors, responsible for the Society's work in large territorial "provinces," such as the upper and lower Rhine in Germany, or Brazil in the New World; and below them were the local superiors, responsible for particular regions or cities, right down to individual colleges and residences. Unlike other religious orders, where local communities enjoyed considerable autonomy and could pick their own leaders, power among the Jesuits flowed strictly from top to bottom: it was the superior general in Rome, not the local members, who appointed the provincial superiors, and they in turn, in close consultation with Rome, appointed the local superiors. The members of each local community were expected to accept these decisions whether they liked them or not, and with very rare exceptions, they did.

The willingness of local Jesuits to submit to the edicts of faraway superiors requires some explanation. After all, the superior general in Rome, capable and dedicated though he was, was often quite ignorant of local conditions, and his directives could be misguided, and even disastrous. Such, for example, was the experience of the French Jesuits in 1594, when they were required to swear allegiance to Henri IV, the new king of France, who had recently converted to Catholicism. The superior general Claudio Acquaviva strictly forbade the Jesuits from taking such an oath, a decision that resulted in their expulsion from Paris and very nearly the end of their activities in France. But even in such extreme situations, when they knew full well that the directions from Rome were misguided and based on a flawed understanding of local conditions, and even when they themselves had to pay the price for their superiors' blunders, the Jesuits obeyed.

The reason was that, for the Jesuits, the principle of "obedience" was not just a practical concession to the requirements of efficient action, but a religious ideal of the highest order. "With all judgment of our own put aside, we ought . . . to be obedient to the true Spouse of Christ our Lord, which is our Holy Mother the hierarchical Church," wrote Ignatius in *The Spiritual Exercises*. This obedience extends not only to actions, but also to opinions and even sense perceptions. "To keep ourselves right in all things," Ignatius wrote, "we ought to hold fast to this principle: What I see as white I will believe to be black if the hierarchical Church thus determines it."

Modern readers might understandably associate such absolute obedience to a ruling hierarchy with the totalitarian regimes that have darkened the history of the twentieth century. Indeed, the requirement that one see white as black if so ordered brings to mind George Orwell's *1984*, in which Winston is required to see four fingers as five in order to prove his loyalty to Big Brother. But there is an important difference: Winston, in *1984*, is being tortured, and is forced to accept Big Brother's supremacy against his will. To the Jesuits, obedience was a high ideal, and its attainment entirely voluntary. Obeying a superior's order, Ignatius wrote, was not an act of abject submission, but a positive reaffirmation of the Society's mission and one's role within it. It followed that although disciplinary measures such as reprimands and even expulsion did exist in the Society of Jesus, they were rarely used in practice. Those who had undergone the rigorous training regimen to become formed Jesuits rarely required such measures to remind them of the value of obedience. Ultimately, Ignatius wrote, "all authority is derived from God," and consequently, obeying the commands of a superior should be immediate and willing, "as if it were coming from Christ our Savior."

In the broadest sense, imposing order on chaos was the Society's core mission, both in its internal arrangements and in its engagement with the world. This was already in evidence in *The Spiritual Exercises*, which transforms an ineffable mystical experience into something like an orderly course of study. It is evident also in Ignatius's *Constitutions*, which provides detailed systematic directions for the running of the Society, and ultimately in the *Ratio studiorum*, the document that outlined in fine detail what must be taught in Jesuit colleges, how, and by whom. Even in their personal lives the Jesuits held to a code of strict orderliness: "Whoever has studied the Jesuits' regimen must be struck by the frequent emphasis on tidiness and order," noted one early twentieth-century historian of the Jesuits. Neatness, cleanliness, and order in both personal quarters and the communal household were "an absolute requirement." Most of all it was expressed in the clear hierarchy of the Society, in which each member was assigned a precise and uncontested place. It was this ability to impose order on chaos that made the Society such an effective instrument in the fight to defeat Protestantism and reestablish the power and prestige of the Church hierarchy.

THE JESUITS STRIKE BACK

Highly educated and fanatically devoted to the cause of the Church and the Pope, the Jesuits were a spiritual army such as Europe had never seen. For the popes, they were a weapon without equal in the former's struggle to impose the authority and the teachings of the Church upon a turbulent and skeptical world, and the popes did not hesitate to make good use of that weapon. From the beginning, the Jesuits were sent on the road to shore up the faith in regions where it was under attack. Pierre Favre, Ignatius's early companion from Paris, was the first Jesuit to work in Germany. The best hope for the Roman Church, Favre surmised, was to strengthen the people's attachment to the traditional holy rites and services: "If the heretics should see in the churches the practice of frequent Communion, with the faithful receiving their Strength and their Life . . . not one of them would dare to preach the Zwinglian doctrine of the Holy Eucharist." He traveled the country, visiting parishes, preaching to large gatherings, and reviving the old communal traditions of the Church.

Favre died in 1546, but two other outstanding Jesuits stepped into the breach: first the Spaniard Jerónimo Nadal, and then Peter Canisius, the "second apostle" of Germany. From the 1540s to the 1560s, Canisius logged approximately twenty thousand miles on the roads of Austria,

Bohemia, Germany, Switzerland, and Italy. Going beyond preaching and the organizational work of reviving parish life, he issued a steady stream of popular books instructing both priests and their flock in correct Catholic doctrines and practices. The results that he and other Jesuits achieved were nothing short of dramatic: The priests at the Jesuit church in Vienna, for example, heard seven hundred confessions in Easter of 1560, but nine years later, the number had grown to three thousand. Similarly in Cologne, in 1576, fifteen thousand worshippers received Holy Communion at the Jesuit chapel, but only five years later the number had tripled, to forty-five thousand. Here was proof of the Jesuits' prowess at reviving Catholic life in lands that were poised for a Protestant takeover.

Jesuits served as the engines of the Catholic revival in other capacities. Some, such as Francisco Suárez, were outstanding formal theologians, who set down the doctrines of the Church and could more than hold their own in debate with their Protestant critics. Others, such as Diego Laynez and Antonio Possevino, served as personal emissaries of the Pope on important diplomatic missions, and still others, such as Robert Bellarmine, combined the two roles as papal theologians and counselors. Some, such as François de la Chaise, personal confessor of Louis XIV and namesake of the famous Père Lachaise cemetery in Paris, provided moral guidance and spiritual solace to European royalty. Others yet, such as the Englishman Edmund Campion, were sent on secret missions to their Protestant homelands to nurture the flame of Catholicism, at enormous risk to themselves. In all these roles the Jesuits proved themselves outstanding religious warriors: learned and often brilliant, skilled, energetic, and zealously devoted to the cause of Church and Pope.

AN EMPIRE OF LEARNING

But while the Jesuits were successful in all these endeavors, it was in one area in particular that they were truly without peer: education. Significantly, apart from training new members, Ignatius did not initially consider education to be a primary focus of his Society. His vision saw the Jesuits as itinerant priests, ready to pack up at a moment's notice and travel to the four corners of the earth at the behest of the Pope or their

superiors, and consequently unsuitable to run schools. But when Francis
Borgia founded the first Jesuit college, in Gandia, Spain, in 1545, the
leading citizens of the town besieged him with requests to allow their
sons to be educated there. Borgia turned to Ignatius, who, sensing an
opportunity to further the cause of Catholic revival, gave his consent. By
1548 the college of Gandia had opened up to the town's youth.

The experience in Gandia set the trend for other institutions. The
year 1548 also saw the opening of the college of Messina, in Sicily, the
first Jesuit institution devoted primarily to educating secular students.
To oversee its founding, Ignatius dispatched a number of his most trusted
subordinates, including Nadal and Canisius, who made Messina into a
model for future colleges. Following Ignatius's instructions, the curriculum included an intensive course in Latin, the classical authors, and
philosophy guided by the writings of Aristotle. At the top of the hierarchy of learning was theology, the "Queen of the sciences," which had
final say on all matters of true knowledge. The faculty at Messina, led by
Nadal, worked to make this broad program of instruction into a systematic and orderly curriculum and issued several proposals for an "order of
studies," or, in its more familiar Latin form, *ratio studiorum*. After under-
going many revisions and several drafts, the *Ratio studiorum* was formally approved in 1599 by the Society's general congregation and became
the blueprint for Jesuit teaching everywhere.

Following these early successes, demand for Jesuit colleges exploded
across Catholic Europe. In towns and cities large and small, ruling
princes, local bishops, and prominent citizens appealed to the Society to
found colleges in their communities. Recognizing the value of education
for spreading the teachings of the Church, Ignatius chose to embrace
this new Jesuit mission, and called for the establishment of Jesuit institutions across Europe. By the time of his death in 1556, there were already 33 Jesuit colleges, and the demand only kept growing: 144 colleges
in 1579, 444 colleges plus 100 seminaries and schools in 1626, and 669
colleges plus 176 seminaries and schools in 1749. Most were in Europe,
but not all. Jesuit colleges could be found as far east as Nagasaki, Japan,
and as far west as Lima, Peru. It was truly a world-encompassing educational system, on a scale the world had never seen before, or has, for that
matter, since.

At the center of this great educational network was the Roman College, known universally as the Collegio Romano. Founded in 1551, it was initially housed in various modest locations around Rome. Pope Gregory XIII (1572–85), an admirer and supporter of the Jesuits, decided to give their flagship institution a more fitting home. He expropriated two city blocks near the main thoroughfare of the Via del Corso and commissioned the renowned architect Bartolommeo Ammannati to design a suitable headquarters for the Jesuit educational system. The result was a large and impressive, though hardly ostentatious, palazzo that reflected the power and prestige of the Society of Jesus but also the seriousness of its mission and down-to-earth pragmatism. The College moved into its new home in 1584, and it was there, nearly half a century later, that the Revisors General met to rule on the fate of infinitesimals. It would remain there, in the Piazza del Collegio Romano, almost continuously for the next three centuries.

The simple name of the Roman College, no different from a Jesuit college in any other city, suggests that it was meant to serve the young men of Rome just as, say, the "Cologne College" was created to educate the youth of that city. But this is misleading. Although educating the Roman elite was indeed part of the college's mission, it was also, from its inception, a model and intellectual beacon for the other colleges in the system. Only the most accomplished Jesuit scholars were summoned to Rome to serve as professors at the Collegio, which brought together under one roof the greatest luminaries of the order. Mathematicians Christopher Clavius and Christoph Grienberger, natural philosophers Athanasius Kircher and Roger Boscovich, theologians Francisco Suárez and Robert Bellarmine, and many others—nearly all, in fact, of the leading Jesuit intellectuals—taught at the Collegio Romano. In keeping with the Society's hierarchical practices, the Roman faculty had the authority to set the curriculum of the provincial colleges and determine what would and would not be taught in Jesuit schools. Just as the order's superior general ruled over each and every individual Jesuit, so the Roman College ruled over all the hundreds of Jesuit colleges worldwide.

It is not difficult to see why aristocrats and wealthy commoners across Catholic Europe clamored for the establishment of Jesuit colleges in their towns. Traditional parochial schools were of dubious quality, and student life at the great universities was reputedly dissolute and immoral, and little concerned with actual studies. The Jesuits offered something else altogether: a rigorous and demanding curriculum taught by highly qualified teachers and regularly updated by the luminaries of the Collegio Romano. And whereas university students were free to indulge in a life of drunken debauchery, the students in the Jesuit colleges

The Collegio Romano, designed by Bartolommeo Ammannati, as it appears today. The building currently houses a state school.
(Alinari / Art Resource, NY)

were closely supervised and filled their days with study and prayer. An aristocrat or merchant who sent his son to a Jesuit school was confident that the boy would be immeasurably bettered, both intellectually and morally.

The long list of distinguished alumni of Jesuit colleges fully bears out this assessment. In addition to the leading Jesuits themselves, graduates include royalty such as Emperor Ferdinand II (1620–37), statesmen such as Cardinal Richelieu, humanists such as Justus Lipsius, and philosophers and scientists such as René Descartes and Marin Mersenne. Jesuit education, as even enemies of the Society acknowledged, was simply the best available in all Christendom. Even Francis Bacon, Lord Chancellor of England, and no friend of the Jesuits, ruefully remarked, "Talis quus sis, utinam noster esses" ("you are so good, would that you were ours").

Bacon had good reason to rue the Jesuits' educational excellence. For, of all the services the Society of Jesus offered the Papacy in its struggle against Protestantism, none proved more powerful or more effective than the colleges. Wherever one was established, it became a center of Catholic life and a living demonstration of what the Roman Church could accomplish. Rare was the Lutheran or Calvinist school that could match the Jesuits for sheer educational quality, or compete with them in attracting the sons of the lay elite. Once they had them in their care, the Jesuits spent years imparting Catholic teachings to their charges, complete with learned and authoritative refutations of Protestant doctrines. Inevitably the students became imbued with the Jesuit devotion to the Papacy, and with the Jesuit spirit of dedication and sacrifice for the cause of the Church and its hierarchy. With hundreds of such colleges across Europe, and with hundreds and sometimes thousands of students enrolled in each one, the Jesuit educational system turned out generations of well-educated and devoted Catholics who would ultimately take up leadership positions in their communities. In effect, as the chief educators of the Catholic elite, the Jesuits ensured the survival, as well as the revival, of the Roman Church in large parts of Europe.

The impact of the Jesuit colleges was unmistakable. The first Jesuit college in the Holy Roman Empire was founded in Cologne in 1556, at a time when the empire appeared on the verge of succumbing to the Lutheran surge. But with the college in place, Cologne became a Catholic stronghold, and a base for future expansion of Jesuit activities. In the following decades, with strong support from the ruling Wittelsbach and Habsburg families, the Jesuits founded dozens of colleges in Bavaria and Austria, and took over the administration of existing universities. They even went so far as to found a special school in Rome dedicated to training the promising young Germans for positions as high Church officials. Upon completing their studies, the graduates of this "Collegium Germanicum" returned home, where they became bishops and archbishops, and the backbone of the Catholic revival in Germany. In the Low Countries, too, the Jesuits were exceedingly active: when the northern provinces turned to Protestantism and took up arms against their Habsburg sovereign, the Jesuits helped make the southern provinces into a Catholic bastion. Thanks in great part to their efforts, the region was saved for the Catholic Church, acquired its own separate identity, and ultimately gained independence as the modern state of Belgium.

Much like Germany, sixteenth-century Poland seemed well on its way to accepting one form or another of Protestantism when Catholic noblemen invited the Jesuits to open their colleges there in the 1560s. They soon gained the trust and support of the Polish royal family, which helped the Jesuits expand from five colleges in 1576 to thirty-two colleges by 1648. The Jesuits became the educators of the Polish ruling class, both the rural aristocracy and the urban elite, while in Rome they educated a devoted cadre of priests who returned to Poland to take up the leadership of the Church. So close were the Jesuits to the Polish monarchs, that King Sigismund III (1587–1632) was known as the "Jesuit King" and his son Jan II Kazimierz (1648–68) was a member of the order and a cardinal before assuming the throne. Poland was transformed: a nation that had previously prided itself on religious tolerance, and had opened its churches and parishes to the reformers, became the devout Catholic land we still recognize today. In Poland as elsewhere, the Jesuit intervention proved decisive.

The upright disciples of Ignatius accomplished what the worldly Renaissance popes could not: they arrested the seemingly unstoppable progress of Protestantism across Europe and revived the power and prestige of the Roman Church. Wherever the Society raised its standard and opened its colleges, a new energy of spiritual devotion and purposefulness of action infused the old Church and inspired its followers to make a stand against the heretics. A grateful Pope Gregory XIII acknowledged as much when he addressed the general congregation of the Society in 1581:

> Your holy order . . . is spread throughout the entire world. Anywhere you look you have colleges and houses. You direct kingdoms, provinces, indeed, the whole world. In short, there is this day no single instrument raised up by God against heretics greater than your holy order. It came to the world at the very moment when new errors began to be spread abroad. It is all important therefore . . . that this order increase and prosper from day to day.

ORDER OUT OF CHAOS

The Miracles of St. Ignatius, a massive painting originally intended to grace the altar of Antwerp Cathedral, hangs today at the Kunsthistorisches Museum in Vienna. It is the work of the Flemish painter Peter Paul Rubens (1577–1640), whose modern-day reputation rests largely on his erotic depictions of ample women that challenge our ideal of female beauty. But Rubens was a devout Catholic who attended Mass every morning, and was on intimate terms with the Jesuits in his home city of Antwerp. In 1605, when the Jesuits were campaigning to have their founder canonized, Rubens contributed eighty engravings to a Jesuit hagiography, *The Life of Ignatius*. Four years later, when Ignatius was beatified, placing him one short step from canonization, Rubens was commissioned by the Society to produce several large portraits of the future saint for the Gesù, the Jesuits' home church in Rome, and for the Antwerp Cathedral. In the imposing *Miracles of St. Ignatius*, he accomplished what was probably his greatest masterwork for the order.

The painting brings us to a scene of high drama taking place within a large hall, most likely a church, which is depicted from its vaulted ceiling to its stony floor. At the top, near a brightly lit cupola, floats a band of playful angels and cherubs, who seem to pay no heed to the human chaos beneath them. Indeed, the floor of the church is a scene of pain, fear, and confusion, where a large group of men, women, and several children are caught up in an agonizing frenzy. One man flails about on his back as if in a seizure, while another man, with bloody streaks on his back, tends to him. A disheveled woman, her fists clenched, her face twisted wildly, and her eyes glazed, struggles to get away as two men try to support her. A gray-haired man, only his head visible, gazes up in desperation, his face twisted in a mask of horror. The rest, those who have not been overcome themselves, look upward in a tortured mixture of supplication and hope: can they be saved from that which torments them?

The figure that is the object of their gaze is Ignatius himself, standing upright, resplendent in his priestly robes. On his dais, Ignatius is only a few steps above the floor, but he inhabits a completely different realm. Calm and commanding, his right hand raised in benediction, he is performing an exorcism, expelling the evil spirits from the people, bringing peace and order to those afflicted with torment and chaos. Evil demons, on the left side of the painting, have emerged from the people and are fleeing before Ignatius's holiness as one of the angels in heaven bids them an ironic good-bye. Ignatius, though the unchallenged focus of the painting, is not alone: behind him on the raised platform are his followers, a long line of black-robed Jesuits stretching into the distance. Like him, they are calm and somber, surveying the suffering before them. They are Ignatius's army, there to learn from their master, follow his directions, and ultimately take over his mission of turning chaos into order and bringing peace to the afflicted.

For that was indeed the "miracle" of St. Ignatius and his followers. Like no one else, they managed to restore peace and order in a land torn apart by the challenge of the Reformation. In place of heresy and confusion they brought unity and orthodoxy; where the rule of the holy Church was subverted and priests and bishops disowned, they rebuilt that grand old edifice and reestablished the sway of its hierarchy; where confusion reigned, they restored an unyielding certainty in the truth and rightness of the Roman Church. Their success in doing all this was indeed nothing less than miraculous. The keys to this miracle, as the Jesuits saw it, were simple: truth, hierarchy, and order.

The Jesuits did not believe in plurality of opinion: the truth was absolute. They did not believe in pluralism of power and authority: once the truth is known, all power must flow from those who know and recognize it, and be imposed upon those who do not yet accept it. And they

Peter Paul Rubens, *The Miracles of St. Ignatius*, 1617.
(Erich Lessing / Art Resource, NY)

certainly did not believe in democracy, which allows the expression of different and opposing views and thrives on lively debate and competition for power. The truth has no room for such dissent or challenges. Only absolute authority of God's emissaries and the divine truth they carry, they believed, would allow for peace and harmony to prevail. Such was the worldview of the Jesuits, and they worked hard to implement it within their order, within the Church as a whole, and in the world at large. Structured in a clear hierarchy, *The Miracles of St. Ignatius* puts this entire narrative into visual form. At the top is the realm of divine light and truth; at the bottom are the tormented and confused people. In between are Ignatius and his men: disciplined, calm, and commanding, they expel the demons of strife and confer the light of truth upon the people. Thanks to the Jesuits, peace will prevail.

2

Mathematical Order

TEACHING ORDER

Ignatius of Loyola, founding father of the Society of Jesus, was not enamored of mathematics. As an aristocratic courtier and dashing cavalier
in his early life, he learned to despise the pedantries of scholars and
mathematicians. The ecstatic revelations of his later years led him, if
anything, even farther away from the cold, logical world of numbers and
figures, and his university studies in Barcelona, Alcalá, Salamanca,
and Paris did not apparently include any mathematics. By 1553, when
under his leadership the Jesuits were in the process of launching a
worldwide network of colleges, Ignatius came to see the value of some
mathematical education, writing that the colleges should teach "the
parts of mathematics that a theologian should know." And that, it should
be admitted, was not much.

The low standing of mathematics in the early days of the Jesuit educational system is not, in truth, surprising. The Jesuit colleges, after all,
had a very specific and urgent goal, very different from the aims of their
modern successors: to stop the spread of Protestantism and reestablish
the prestige and authority of the Catholic Church. As Ignatius's lieuten-
ant, Juan de Polanco, explained in a 1655 letter, in the Society's colleges
"men of those nations" where the true faith is threatened "are taught
with example and sound doctrine . . . to keep what remains, and restore

what was lost, of the Christian religion." A remote and abstract subject such as mathematics had little to contribute to this mission.

The goal of reversing the progress of the Reformation, however, did not mean that the Jesuit colleges focused their curriculum exclusively on religious teachings. Ignatius firmly believed that proper religious instruction must be grounded in broader teachings in philosophy, grammar, classical languages, and other humanistic fields, and it was also essential that the colleges live up to their promise of providing broad and up-to-date education. Otherwise, the local elites would turn elsewhere for the education of their sons, which would spell disaster for the order's spiritual mission. As Jerónimo Nadal put it in 1567, "For us lessons and scholarly exercises are a sort of hook with which we fish for souls."

The "hook" that Ignatius recommended included the languages that might be required in order to read the ancient masters: Latin, Greek, and Hebrew, but also, in some colleges, Chaldee, Arabic, and Hindi. In philosophy, he ruled that the colleges would follow the teachings of the ancient Greek philosopher Aristotle, by far the most influential philosopher in the West ever since his writings had been translated into Latin in the twelfth century. His corpus of writings, covering fields as diverse as logic, biology, ethics, politics, physics, and astronomy, was the most comprehensive then known, and was accepted as authoritative by the majority of European scholars. It was therefore easy for Ignatius, who had studied Aristotle at the universities, to rely on him in setting the curriculum for the Society's colleges. In theology, Ignatius decreed, the Society would follow St. Thomas Aquinas, the thirteenth-century Dominican who reconciled the teachings of Aristotle and the Church. "The Angelic Doctor," as Aquinas was known, became after his death the most authoritative theologian in the West, and Ignatius regarded him as well-nigh infallible. Since Thomism (as Aquinas's theology was known) relied heavily on Aristotelian philosophy, it was essential that students in the Jesuit colleges be immersed in Aristotle before they fully engaged in religious study.

But if the curriculum of the Jesuit colleges was diverse and wide-ranging, it was also rigorous, clearly ordered, and hierarchical. The relative value of the different disciplines was never in doubt: At the top was theology, comprised of the infallible teachings of the Catholic Church.

Below it was philosophy, both moral and natural, which taught truths about the natural and human world and might be required in order to understand religious teachings. And below philosophy were the ancillary fields such as languages and mathematics, which did not deal with truth themselves but could prove useful in understanding the higher disciplines. Here, as elsewhere in the Jesuit world, order prevailed. Each field had its place in the grand scheme of the disciplines. The truths of theology were the highest, and no philosophical doctrine, even if supported by the authority of Aristotle himself, could ever contradict a theological truth. Mathematical sciences ranked lower still, and their results did not even qualify as truth, but only as hypotheses. It was a seamless hierarchy of knowledge in which Thomist theology reigned supreme.

The clear order of the disciplines at the Jesuit colleges contrasted well with the offerings at the universities of the time, where studies were often haphazard and students typically attended unrelated lectures. Many students lost their way in this unstructured maze. The Jesuits, in contrast, offered a clear sequence of learning, beginning with languages and the many branches of Aristotelian philosophy, then moving on to theology. Along with the regulated and orderly life of the colleges and the upstanding moral example of the Jesuit instructors, this rigid progression kept the students on track and away from the temptations that afflicted their peers.

But the hierarchy of truth was, for the Jesuits, more than a pedagogical device. It reflected their unyielding faith that a clear and undisputed hierarchy was essential for reconstituting the godly order lost in the Reformation. It governed society itself, and it governed the Church, from the Pope to the lay congregation. Hierarchy, the Jesuits believed, must prevail in the world if heresy were to be defeated and if truth were to triumph over error. After all, was not the scourge of the Reformation itself the result of a breakdown in the proper order of knowledge? Did not Luther, a mere monk, dare to challenge the authority of the Pope himself? Did not Luther and, later, Zwingli, Calvin, and others posit their own novel theologies in opposition to the authoritative teachings of the Church? And what was the result? Chaos and confusion, in which the single authoritative voice of the Roman Church was drowned by a cacophony of competing voices. It seemed obvious to the Jesuits that the collapse of the ancient unity of Christendom, and the chaos that followed, were the direct result of the collapse in the proper order of knowledge. Only by preserving this strict hierarchy of knowledge would truth prevail and heresy be defeated.

Since truth, for the Jesuits, was unchanging, and eternal, and founded on the authority of the Church, then novelty and innovation posed an unacceptable risk, and must be fervently resisted. "One should not be drawn to new opinions, that is, those that one has discovered," warned theologian Benito Pereira of the Collegio Romano in 1564. Instead, one must "adhere to the old and generally accepted opinions . . . and follow the true and sound doctrine." Two decades later, General Acquaviva exhorted his cohorts to avoid not only innovation, but also having "anyone suspect us of trying to create something new." Innovation, so prized today, was regarded with deep suspicion by the Jesuits.

Legem impone subactis—impose your rule upon the subjects!—was the motto of the Accademia Parthenia at the Collegio Romano, open to those at the college who were exceptionally devoted to the Jesuit ideals and way of life. Accompanying the motto was its equally transparent coat of arms, known as an "impresse." At the top, seated on a throne, is the female figure of Theology. Flanking her, on a lower plane, are her servants Philosophy and Mathematics, reclining and awaiting her command. And so it was in the Society's schools, where theology reigned as the "queen of the sciences" and imposed her rule upon subordinate subjects. It is a system of knowledge that seems alien to us today, even stifling, designed as it was to establish absolute truths and quash dissent. But the Jesuits believed that the purpose of education was not to encourage the free exchange of ideas, but to inculcate certain truths. And in that, they were undeniably successful.

AN UNAPPRECIATED MAN

So things stood in the first decades of the Society of Jesus, when mathematics, if addressed at all, was taught only to the extent that it was useful for other, higher disciplines. And so things would likely have remained were it not for the work of one man who made it his life's mission to bring mathematics to the center of the Jesuit curriculum. It was thanks to his efforts that, by the dawn of the seventeenth century, the

Jesuits had become not only skilled teachers of mathematics, but also leading scholars in the field, numbering among their own some of the most prominent mathematicians in all Europe. His name was Christopher Clavius.

Little is known of Clavius's early years—even his true birth name remains in doubt—but we do know that he was born on March 25, 1538, in the city of Bamberg, in the south German province of Franconia. As the seat of a Catholic prince-bishop, but surrounded by the Protestant territories of Nuremberg, Hesse, and Saxony, Bamberg was on the front lines of the struggle for the soul of the Holy Roman Empire. It was cities such as Bamberg that were targeted by the Jesuit Peter Canisius as he barnstormed across the empire, reviving the sagging spirits of the faithful and exhorting them to take a stand against the encroaching Protestant tide. It is easy to imagine the young Clavius attending one of Canisius's giant Masses at Bamberg Cathedral and being moved by his fiery preaching, but we don't know this for a fact. What we do know is that, in 1555, as his home city was fending off the forces of the Protestant margrave Albert Alcibiades, Clavius was in Rome. On April 12 he was received as a novice into the Society of Jesus by Ignatius of Loyola himself.

Clavius was just seventeen when he joined the Society, but he was thirty-seven by the time he professed his final solemn vows. Even considering the lengthy and rigorous Jesuit training regimen, twenty years is an unusually long time for a bright young man to rise from novice to fully formed Jesuit, especially for one who was recognized for his promise early on, and who would ultimately become one of the most famous Jesuits of the age. But it may have had something to do with the fact that Clavius spent much of that time campaigning inside the Society for an unpopular cause: raising the status of mathematics in the Jesuit hierarchy of knowledge and improving the teaching of it in the order's schools. Some Jesuits, such as Benito Pereira, who was Clavius's colleague at the Collegio Romano, vigorously opposed him. Nevertheless, by the time Clavius joined the ranks of the "professed" Jesuits in 1575, he was well on his way to winning the fight.

Clavius spent only a year in Rome after being admitted to the Society as a novice before he was sent off to the Jesuit house in Coimbra,

Portugal. Unlike the secluded monasteries of traditional orders such as the Benedictines, such "houses" (or "residences") were located in the heart of the city or town. There, the local Jesuits lived as a tight-knit community under an appointed superior, and emerged daily to conduct their activities in the broader community. Little is known about the four years Clavius spent in Coimbra in his late teens and early twenties, but they were undoubtedly formative ones. The city was famous in those days as the seat of an ancient university, whose most celebrated resident was Pedro Nuñez, one of the great mathematicians and astronomers of the age. There is no direct evidence that Clavius studied under Nuñez, but the mathematician Bernardino Baldi (1553–1617), who wrote a short biography of Clavius, does mention that the two knew each other. To be sure, given the young German's interests and the small size of the University of Coimbra, it is hard to imagine that Clavius and Nuñez did not meet. But, for the most part, according to Baldi, Clavius was self-taught, gaining his knowledge of mathematics through his own careful study of classical mathematical texts.

When Clavius was recalled to Rome in 1560, it was to continue his study of philosophy and theology, and to teach mathematics. In 1563 he was lecturing on mathematics at the Collegio Romano, and around 1565, when thirty years old, he became a professor of mathematics, a position that he would hold more or less continuously until his death forty-seven years later. Up to this point, Clavius's career was respectable but hardly remarkable. Although recognized by his superiors for his mathematical abilities, he was nevertheless just a young faculty member toiling in obscurity among colleagues who did not much respect his field of expertise. Even years later he was still fighting for the right of the mathematics professor to take part in public ceremonies and disputations along with his colleagues, a complaint that suggests that this was not usually done. And despite holding a chair at the Society's flagship college, he was excluded for years from the ranks of the "professed," which tells us all we need to know about his status in the rigid hierarchy of the order.

But sometime between 1572 and 1575, more than a decade after his return from the provinces, Clavius's career took a dramatic turn. The newly elected Pope Gregory XIII assembled a distinguished commission to deal with an issue that had troubled the Church for centuries: calendar reform. As technical adviser to the commission the Pope selected the young Jesuit professor of the Collegio Romano who was making a name for himself as an expert on mathematical and astronomical matters. The appointment was unquestionably a great honor for Clavius, putting him at the center of one of the most ambitious projects the Church had undertaken. It also made him the official representative of the Jesuits in a high-ranking panel of the Church, whose recommendations would be known to all and would be scrutinized by scholars across Europe. Placed in such a visible position, Clavius was expected to bring honor and distinction to the Society, and enhance its prestige in the papal court. It was a difficult and even risky proposition for a young and rather obscure professor of mathematics. But Clavius and his cause had been waiting for just such an opportunity.

ORDERING THE UNIVERSE

The problem the commission was called to address had been in the making for more than twelve hundred years. Back in the year 325 CE the Council of Nicea had determined that Easter should be celebrated on the first full moon after the vernal equinox, which, according to the council, fell on March 21. Unfortunately the Julian calendar that was used at the time did not quite match the true length of the solar year— the time that it takes the sun to return to the exact same spot in the sky. Whereas the Julian year is 365 days and 6 hours, the true solar year is almost exactly 11 minutes shorter. Such a minuscule discrepancy does not matter from one year to the next, or even over a person's lifetime, but an error of 11 minutes repeated more than 1,200 times does add up. By the 1570s the date of the vernal equinox had slipped to March 11, and the date of Easter, the most important feast in the Christian calendar, had slipped with it. If nothing were done to correct the problem, the error would continue to grow, and Easter continue to slip. The lunar calendar, meanwhile, which is used to calculate when a full moon will occur, had a comparable problem, slipping one day every 310 years. By the sixteenth century the full moon would appear four days after the date predicted by the calendar.

All this was unacceptable: not only was the date of Easter at stake, but the entire religious calendar of feasts and saints' days, not to mention the seasonal and agricultural calendar, was thrown into disarray. Already in the thirteenth century the English philosopher Roger Bacon had complained that the calendar was "intolerable to all the wise, horrible to all astronomers, and ridiculed by all computists." In truth, the very sense of time and its regularities was perturbed for all Christendom, and the Church, guardian of the sacred rhythms of life, was called to take action. Several Church councils, beginning with the Council of Constance (1414–18), tried to address the problem, but nothing came of these efforts. Finally the Council of Trent, which met periodically in the northern Italian town of Trento between 1545 and 1563, ordered that a special commission be convened for the express purpose of reforming the calendar. Around a decade later, the newly elected Pope Gregory XIII finally acted on the council's decree.

The task of the commission, of which Clavius was a member, was complex. First it had to determine the exact size of the errors in the Julian and lunar calendars. Then it had to produce new lunar tables that would accurately predict the future phases of the moon. Finally, it had to correct for the cumulative slippage that had already taken place, and propose a new calendar that would prevent the error from recurring. In 1577 the commission sent a "compendium" of proposed changes to leading Catholic scholars, soliciting comments and suggestions. After reviewing and sorting through the many responses, the commission was particularly impressed with the elegant and simple proposals of the Calabrian doctor Aloysius Lilius. In September of 1580, when the commission presented its conclusions to the Pope, it based its recommendations largely on Lilius's suggestions.

The first recommendation was for an immediate one-time correction to the calendar that would eliminate ten days. To prevent the problem from reemerging in future centuries, the commission also proposed a permanent adjustment to the Julian calendar: As before, every year that was divisible by 4 would be a leap year, lasting 366 days instead of 365. But unlike the old calendar, years that were divisible by 100 (e.g., 1800, 1900) would be standard 365-day years, with the exception that years that were divisible by 400 would remain leap years. The combined effect would be to reduce the average length of a year by 10 minutes and 48 seconds, effectively synchronizing the calendar year with the solar year.

Henceforth the vernal equinox would always fall on March 21. In February 1582, in the the papal bull "Inter gravissimas," the Pope made it official: Accepting the commission's recommendations, he decreed that Thursday, October 4, of that year would be followed by Friday, October 15, making 1582 the only 355-day year on record. He also instituted the calendar devised by Clavius and his colleagues, the one still used around the world today. It is known, appropriately, as the Gregorian calendar.

Throughout this entire process Clavius's astronomical and mathematical expertise was indispensable. It had been his job to present the most up-to-date astronomical calculations to his less technically adept colleagues on the committee. He also undoubtedly played a leading role in recalculating the phases of the moon and in vetting various scholars' proposals for calendar reform. Through it all, he proved himself not only an excellent mathematician and astronomer, but also someone who could effectively navigate the intricate politics of the papal court. In later years, when the other members of the commission had returned to their regular occupations, Clavius would emerge as the public spokesman for the new system, publishing a six-hundred-page "explanation" of the new calendar and taking on its vociferous critics. The obscure and underappreciated professor at the Collegio Romano had become a leading mathematician, a spokesperson for the rising mathematical sciences, a "professed" Jesuit, and a public face of the order. He would never look back.

A MATHEMATICAL VICTORY

The Gregorian reform of the calendar was a spectacular triumph for the Catholic Church in the dark years of its struggle with the Protestant "heretics." Here was the Pope exercising his universal authority to correct a problem that had troubled all Christians for more than a millennium. In a display of near-godlike power, the Pope transformed the year, the religious festivals, and the seasons for millions across the globe. A bas-relief on a monument to Pope Gregory XIII in St. Peter's Basilica in Rome by the sculptor Camillo Rusconi shows members of the calendar reform commission, with Clavius (according to Jesuit tradition) kneeling in the middle. He is presenting the new calendar to the Pope, who is seated on his throne with his arms spread wide, gesturing toward a globe, as if it were once again his and no one else's. While the Pope's Protestant enemies certainly counted among their number scholars as learned and accomplished as Clavius and his fellow commission members, no Protestant prince or clergyman could become, as the Pope had, a master of time itself.

Protestants had no choice but to acknowledge the binding power of the Pope's proclamation, and his unrivaled ability to reorder the universe. Just how troubling this was to them can be glimpsed in *Ignatius His Conclave*, an anti-Jesuit satire by the English poet and clergyman John Donne dating from 1611. Ignatius, in Donne's depiction, resides in hell with his associates, Clavius among them. "Our Clavius," Ignatius proclaims, is to be honored

> for the great pains . . . which he tooke in the *Gregorian Calendar*, by which both the peace of the Church and the Civill businesses have been egregiously troubled: nor hath heaven itself escaped his violence, but hath ever since obeied his appointments: so that S. Stephen, John Baptist, & all the rest, which have been commanded to work miracles at certaine appointed daies . . . do not now attend till the day come, as they were accustomed, but are awaked ten daies sooner, and constrained by him to come downe from heaven to do that businesse.

The biting satire does not mask the anti-Catholic Donne's sincere consternation at having to succumb to the Pope's reordering of both religious and civil time.

Protestant princes were forced into an unpleasant choice: they could either accept the Gregorian calendar, and thereby implicitly acknowledge the universal authority of the Pope, or they could reject it, and knowingly retain an embarrassingly erroneous calendar. Feeling cornered, they reacted with understandable confusion. Queen Elizabeth I of England at first announced that she would go along with the reform, only to backtrack in the face of opposition from the Church of England. The Gregorian reform would not come to the British Isles until 1752. The Dutch Republic split, with some provinces adopting the reform immediately while others retained the Julian calendar until 1700; and Sweden went back and forth between the two calendars until finally settling on the Gregorian in 1753. Farther east, the Russian Orthodox Church, whose quarrel with the Pope predated Luther's by seven hundred years, also held on to the Julian calendar, until the Bolsheviks, who could not be suspected of being papal agents, imposed the reformed calendar in 1918. The last European nation to adopt the Gregorian calendar was Greece, in 1923, almost three and a half centuries after Clavius and his associates completed their work. By enacting a calendar that effectively took over the world, the Roman Church exhibited commanding authority, whereas its rivals showed only weakness and confusion, and the inherent limitations of their national churches.

The calendar reform was precisely the kind of triumph the Society of Jesus was laboring to achieve. Here was a perfect example of the Catholic Church imposing truth, order, and regularity upon an unruly world. Like St. Ignatius in Rubens's masterpiece, Pope Gregory was bringing the light of universal truth to the people, who had long suffered in darkness and confusion. The reaction to the reforms confirmed this: wherever the Pope's command was law, order, peace, and truth prevailed; wherever heretics and schismatics ruled, error, confusion, and strife persisted. Nothing could better illustrate the justice of the approach that was at the core of the Jesuit worldview. Here, the Jesuits believed, was a template for the ultimate triumph of the Roman Church.

The decisive victory of the Roman Church in the matter of the calendar seemed all the more striking when compared to the stalemate that prevailed in other areas of theological dispute. Catholics, for example, believed that the grace of God was bestowed upon sinners only through the holy Church and its sacraments, enacted by an ordained priest. Protestants, conversely, believed in a "priesthood of all believers," meaning that God would bestow his grace directly upon them. Catholics believed that Christ was physically present in the bread and wine during the sacrament of the Mass. Protestants believed that Christ was either present everywhere (Luther) or that the Mass was a mere commemoration of his sufferings (Zwingli). Catholics believed that God would take into account a man's good works in this world in determining whether he would be saved or lost. Protestants believed that only faith and divine grace mattered. Catholics believed that the Bible required interpretation by the hierarchy and the traditions of the Church. Protestants believed that the Bible was a clear guide for righteous behavior, accessible to anyone. And so on. What these arguments had (and still have) in common is that they are entirely inconclusive. From Luther's day to ours, neither side has given an inch, or has seen any reason to.

To be sure, advocates on both sides engaged in passionate, and often violent, debate. They published crude caricatures of each other, depicting Luther as the Devil's emissary, or the Pope as anti-Christ, and disseminated them as broadly as the new technology of the printing press allowed. They published popular pamphlets denouncing each other's doctrines as heretical, and catechisms detailing the fundamentals of each faith. They authored learned treatises such as Calvin's *Institutes of the Christian Religion* or the Jesuit Francisco Suárez's *Disputationes metaphysicae*, and they engaged, occasionally, in formal debate, as Luther had with Eck in 1519. But despite the effort, time, and resources invested in these battles, neither side was able to impose its position upon the other. What a contrast between this bog of indecision and the glorious, clear-cut victory afforded the Roman Church by the reform of the calendar! If only the secret of the calendrical triumph could be infused into those other fields, the ultimate victory of Pope and Church would be assured.

Clavius believed that he knew what this secret was: mathematics. Theological and philosophical disputes could rage forever, he believed, because there was no universally accepted way to decide who was right and who was wrong. Even when one side possessed the absolute truth (as Clavius believed it did), and the other nothing but error, the adherents of error could still refuse to accept the truth. But mathematics was different: with mathematics, the truth forces itself upon its audience whether they like it or not. One could dispute the Catholic doctrine of the sacraments, but one could not deny the Pythagorean theorem; and no one could challenge the correctness of the new calendar, based as it was on detailed mathematical calculations. Here, Clavius believed, was a key to the ultimate triumph of the Church.

THE CERTAINTY OF MATHEMATICS

Clavius elaborated his views on mathematics in an essay that he attached to his edition of Euclid, which first came out in 1574, just as the commission on the calendar was getting down to work. Entitled simply "In disciplinas mathematicas prolegomena" ("Introductory Essay on the Mathematical Sciences"), it is in fact a passionate appeal for recognition of the power of the mathematical sciences and their superiority over other disciplines. If "the nobility and the excellence of a science is to be judged by the certainty of the demonstrations that it uses," Clavius wrote, then "without a doubt the mathematical disciplines have the first place among all others": "They demonstrate everything in which they see a dispute by the strongest reasons, and they confirm it in such a way that they engender true knowledge in the minds of the hearers, and completely remove any doubt." Mathematics, in other words, imposes itself on the minds of its hearers and compels even the most recalcitrant among them to accept its truths.

"The theorems of Euclid," he continues, "and the rest of the mathematicians,"

> still today as for many years past, retain in the schools their true purity, their real certitude, and their strong and firm demonstrations . . . And thus so much do the mathematical disciplines desire, esteem, and foster the truth that they reject not only whatever is false, but even anything merely probable, and they admit nothing that does not lend support and corroboration to the most certain demonstrations.

But the case is very different with the other so-called "sciences." Here, Clavius argues, the intellect deals with a "multitude of opinions" and a "variety of views on the truth of the conclusions that are being assessed." The result is that whereas mathematics leads to certainty that ends all debate, other fields leave the mind confused and uncertain. Indeed, Clavius continues, commenting on the inherent inconclusiveness of nonmathematical fields, "how far all this is from mathematics, I think no one admits." "There can be no doubt," he concludes, "but that the first place among the other sciences should be conceded to mathematics."

Rigorous, orderly, and irresistible, mathematics was for Clavius the embodiment of the Jesuit program. By imposing truth and vanquishing error, it established a fixed order and certainty in place of chaos and confusion. It should be remembered, however, that when Clavius is speaking of "mathematics," he has something quite specific in mind. Certainly the arithmetic in use by merchants and traders had its place, as did the emerging new science of algebra, which teaches one how to solve quadratic, cubic, and quartic equations. But the true model of mathematical perfection for Clavius was geometry, as presented in Euclid's great opus *The Elements*. It was the only mathematical field, he believed, that captured the power and truth of the discipline in its most distilled form. When Clavius wished to emphasize the eternal truth of mathematics, he cited "the demonstrations of Euclid," and surely it is no coincidence that of all his textbooks on the many mathematical fields, he chose to append his "Prolegomena" to his edition of Euclid.

Composed around 300 BCE, *The Elements* is arguably the most influential mathematical text in history. But not because it presented new and original results: *The Elements* was based on the work of earlier generations of geometers, and most of its results were likely well known to practicing mathematicians. What was revolutionary about Euclid's work was its systematic and rigorous method. It begins with a series of definitions and postulates that are so simple as to be self-evidently true. According to one definition, "A figure is that which is contained by any boundary or boundaries"; according to one postulate, "all right angles are equal to one another"; and so on. From these seemingly trivial beginnings, Euclid moves step by step to demonstrate increasingly complex results: that the base angles of an isosceles triangle are equal; that in a right triangle, the sum of the squares of the two sides containing the right angle is equal to the square of the third side (the Pythagorean theorem); that in a circle, the angles in the same segment are all equal to one another; and so on. At each step, Euclid does not just argue that his result is plausible or likely, but demonstrates that it is absolutely true and cannot be otherwise. In this manner, layer by layer, Euclid constructs an edifice of mathematical truth, composed of interconnected and unshakably true propositions, each dependent on the ones that precede it. As Clavius points out in the "Prolegomena," it was the sturdiest edifice in the kingdom of knowledge.

For a taste of the Euclidean method, consider Euclid's proof of proposition 32 in book 1: that the sum of the angles of any triangle is equal to two right angles—or, as we would say, 180 degrees. Euclid, at this point, has already proven that when a straight line falls on two parallel lines, it creates the same angles with one parallel line as with the other (book 1, proposition 29). He makes good use of this theorem here:

Proposition 32: In any triangle, if one of the sides be produced, the exterior angle is equal to the two interior and opposite angles, and the three interior angles of the triangle are equal to two right angles.
    Proof:
    Let ABC be a triangle and let one side of it be produced to D. I say that the exterior angle ACD is equal to the two interior and opposite angles CAB, ABC, and the three interior angles of the triangle, ABC, BCA, CAB, are equal to two right angles.

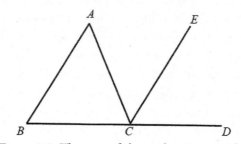

Figure 2.1. The sum of the angles in a triangle.

For let CE be drawn through the point C, parallel to the line AB.
    Then, since AB is parallel to CE and AC falls on both of them, the alternate angles BAC and ACE are equal to one another.
    Again, since AB is parallel to CE, and the straight line BD falls on them, the exterior angle ECD is equal to the interior and opposite angle ABC.
    But the angle ACE was also proved equal to BAC. It follows that the whole angle ACD (composed of ACE and ECD) is equal to the two interior and opposite angles BAC and ABC.
    Let the angle ACB be added to each; it follows that the sum of the angles ACB and ACD are equal to the sum of the interior angles of the triangle, ABC, BCA, CAB.

But since the angles ACB and ACE are equal to two right angles,
it follows that the angles of the triangle, ABC, BCA, CAB are also
equal to two right angles.

Q.E.D.

Euclid's proof here is fundamentally simple: He extends the triangle's
side *BC* to the point *D*, and then draws a parallel line to *AB* through the
opposite corner *C*. Using what he has already proven about the properties
of parallels, he transfers the triangle's angles *A* and *B* to the line *BD* next
to the angle *C*, thus showing that the three angles together combine to
form a straight line—that is, 180 degrees. But even in this simple proof,
all the elements that make Euclid so compelling are clearly present. The
proof is based on previous ones, in this case, the unique properties of parallels; from there, it proceeds systematically, step by step, showing clearly
that each small step is logically correct and necessary; and ultimately, it
arrives at its conclusion, which is absolutely true and universal. Not only
the specific triangle *ABC* has angles that combine to 180 degrees, but
every triangle that ever was, will be, or can be will show the exact same
characteristic. Finally, the proof of proposition 32 and every other Euclid-
can proof is a microcosm of Euclid's geometry as a whole. Just as each
proof is composed of small logical steps, so the proofs themselves are but
small steps in the edifice that is Euclidean geometry. And like each proof
alone, geometry as a whole is universally and eternally true, ordering the
world and governing its structure everywhere and always.

It was clear to Clavius that Euclid's method had succeeded in doing
precisely what the Jesuits were struggling so hard to accomplish: imposing a true, eternal, and unchallengeable order upon a seemingly chaotic
reality. The diverse world we see around us, made of seemingly limitless
shapes, colors, and textures, might appear to us as chaotic and unruly.
But thanks to Euclid, we know better: all this diversity and apparent
chaos is in fact strictly ordered by the eternal and universal truths of
geometry. Antonio Possevino, a Jesuit, papal nuncio (ambassador), and
friend and collaborator of Clavius, made this point in his *Bibliotheca selecta* of 1591, where he argues that

if anyone mentally conceives of God as wisest and as Geometrical architect for all . . . he will understand that the world had been joined by

God from all substances and from the whole of matter; but since he wished to leave nothing discordant and unordered, but to adorn it with ratio, measurement, and number . . . therefore the Craftsman of the world imitated the fairest and eternal exemplar.

God had imposed geometry upon unruly matter, and hence the eternal rules of geometry prevail everywhere and always.

Mathematics, and geometry in particular, was for Clavius an expression of the highest Jesuit ideals and provided a clear road map for the Society as it struggled to build a new Catholic order. In some instances mathematics could be used directly to enhance the power of the Church, as was the case with the reform of the calendar. In other instances mathematics could serve as an ideal model for true knowledge, which the other disciplines could aspire to emulate. Either way, to Clavius one thing was clear: mathematics could no longer languish as an afterthought in the Jesuit empire of learning, but must become a core discipline of the curriculum and a key component in the formation of Jesuits.

CLAVIUS AGAINST THE THEOLOGIANS

The road to establishing mathematics as a core discipline in the Jesuit curriculum was a difficult one. In the first place, Clavius had to deal with those of his colleagues who simply did not believe that mathematics deserved the high position in which he wished to place it. Ignatius, they pointed out, had not placed much stock in mathematics, and the authorities he had prescribed were not particularly favorable to mathematics. Aquinas, Ignatius's chosen theological authority, had only limited use for simple mathematics; Aristotle, the Jesuits' guide in philosophy, assigned mathematics a far smaller role than did his teacher and philosophical rival Plato; and in Aristotelian physics and biology, mathematics played no part at all.

The most outspoken of Clavius's opponents at the Collegio Romano seems to have been the theologian Benito Pereira, the same Jesuit who had proclaimed that one must always "adhere to the old and generally accepted opinions." "My opinion," Pereira declared in 1576, just as Clavius was launching into the project of calendar reform, "is that mathematical disciplines are not proper sciences." The problem with mathematics, according to Pereira, is that its demonstrations are weak, and consequently, it does not produce true knowledge, referred to in the philosophical language of the time as *scientia*. This is because proper demonstrations, according to Aristotle, proceed from true causes—those rooted in the essential nature of the objects discussed. For example, the classic syllogism

> All men are mortal
> Socrates is a man
> Therefore Socrates is mortal

proceeds from the fact that mortality is an essential part of being human. But nothing like this, Pereira argues, exists in mathematics, because mathematical demonstrations do not take into account the essence of things. Instead, they point to complex relations between numbers, lines, figures, etc.—all interesting in themselves, no doubt, but lacking the logical force of a demonstration from true causes. The use of parallel lines, for example, might reveal to us that the sum of the angles of a triangle is equal to two right angles, but the parallel lines did not *cause* this to be true. For all intents and purposes, Pereira suggests, mathematics doesn't even have a true subject matter; it merely draws connections between different properties. If one seeks strong demonstrations, one must turn elsewhere: to the syllogistic demonstrations of Aristotelian physics, which are almost entirely devoid of mathematics.

Not so, retorted Clavius in the "Prolegomena." The subject of mathematics is matter itself, since all mathematics is "immersed" in matter. This, he argues, puts mathematics in a distinguished place in the order of knowledge: both immersed in matter and abstracted from it, mathematics is halfway between physics, which deals only with matter, and metaphysics, which deals with things separated from matter. Mathematics, according to Clavius, should not aspire to equality with metaphysical theology, which deals with things such as the soul and salvation. But it is, nonetheless, clearly in a superior position to the Aristotelian physics favored by Pereira. Whether Clavius won the argument is a matter of opinion. Contemporaries thought that he at least held his own, and that is really all he needed. His rising prestige as the Society's representative on

the calendar commission did more than his logical and rhetorical powers to bolster his arguments, and in any case, he was more interested
in actual pedagogical reform than in abstract philosophical debate.
That is where he directed his fight, and that is where he would ultimately win it.

Clavius laid out his plans for raising the profile of mathematics in the
Society in a document called "Modus quo disciplinas mathematicas in
scholis Societatis possent promoveri" ("The ways in which the mathematical disciplines could be promoted in the Society's schools"), which
he circulated around 1582, shortly after the calendar commission had
completed its work. In order for the program to succeed, he argued, it
was first necessary to raise the prestige of the field in the eyes of the
students. This would require some cooperation from his colleagues, and
he did not hesitate to take direct aim at those he suspected of sabotag-
ing his efforts. He clearly had Pereira and his allies in mind when he
complained that reliable sources had informed him that certain teachers
openly mocked the mathematical sciences. "It will contribute much," he
wrote, to the promotion of mathematics

> if the teachers of philosophy abstain from those questions which do
> not help in the understanding of natural things and very much detract
> from the authority of the mathematical disciplines in the eyes of stu-
> dents, such as those in which they teach that the mathematical sci-
> ences are not sciences [and] do not have demonstrations . . .

"Experience teaches," he added acidly, "that these questions are a great
hindrance to students and of no service to them."

Apart from countering the pernicious influence of hostile colleagues,
Clavius also made positive suggestions for the advancement of mathematics in the Society's schools. First and foremost, he argued, master
teachers must be found "with uncommon erudition and authority," since
without those, students "seem unable to be attracted to the mathematical disciplines." In order to produce a cadre of such capable professors,
Clavius suggested establishing a special school, where the most promising mathematics students in the Jesuit colleges would be sent to pursue
higher studies. Later on, once they took up their regular teaching positions, the graduates of the school "should not be taken up with many

other occupations," but be left to focus on mathematical instruction. To counter antimathematical prejudice, it was extremely important that these highly trained mathematicians be treated by their colleagues with the utmost respect, and invited to take part in public disputations alongside the professors of theology and philosophy. The prestige of mathematics, he explained, required this: "pupils up to now seem almost to have despised these sciences for the simple reason that they think that they are not considered of value and are even useless, since the person who teaches them is never summoned to public acts with the other professors."

Then as now, students were very quick to pick up on which subjects and teachers were valued and which were not, and it was close to impossible for instructors in an undervalued field to get the students to take them seriously. Today it is more likely to be teachers of philosophy and the humanities who complain that their fields are disrespected by instructors in the prestigious mathematical sciences. But even if the roles of the different disciplines are roughly reversed today, the dynamic is still much the same.

THE EUCLIDEAN KEY

Staffing the colleges with qualified teachers was one thing. Giving them something to teach, however, was another, and here again Clavius stepped in with a proposal. Already in 1581 he wrote up a detailed mathematical curriculum, which he called "Ordo servandus in addiscendis disciplinis mathematicis"—literally, "The order to be kept in learning the mathematical disciplines." His complete curriculum consisted of twenty-two lesson sets spread over three years of study, a plan that ultimately proved too ambitious to be generally implemented. In the Jesuit colleges, theology and philosophy still came first. Nevertheless, this did not prevent Clavius from pushing hard to introduce as much of his curriculum as possible into the schools.

The first, most important, and key component of Clavius's curriculum was inevitably Euclidean geometry. Any incoming student would start out by studying the first four books of Euclid, which deal with plane geometry. He would then study the fundamentals of arithmetic, before moving on to astronomy, geography, perspective, and music theory, among others, each according to the accepted authority on the subject: Jordanus de Nemore on arithmetic, Sacrobosco on astronomy, Ptolemy on geography, and so on. But he would return time and again to the greatest master of the mathematical sciences, Euclid, until he had thoroughly mastered the entire thirteen books of *The Elements*. It was a logical sequence of studies, but for Clavius it also represented a deeper ideological commitment. Geometry, being rigorous and hierarchical, was, to the Jesuit, the ideal science. The mathematical sciences that followed—astronomy, geography, perspective, music—were all derived from the truths of geometry, and demonstrated how those truths governed the world. Consequently, Clavius's mathematical curriculum did not just teach the students specific competencies. More important, it demonstrated how absolute eternal truths shape the world and govern it.

Clavius spent much of the final thirty years of his life trying to implement this program. Initially he hoped to incorporate his plan into the Society's *Ratio studiorum*, the master document of Jesuit college education that had been in the works for decades. A draft produced at the Collegio Romano in 1586 so closely follows Clavius's suggestions that he likely authored its mathematics chapter himself. It proposed, for example, that a mathematics professor "who could be Father Clavius" should teach a three-year advanced course in mathematics to instruct future Jesuit teachers in the field. A later draft from 1591 repeated much of the same language, and even warned, as Clavius had, against teachers who would subvert the authority and importance of mathematics. The final version of the *Ratio*, issued in 1599 and officially approved, was drier and shorter than its more florid predecessors, but it, too, accepted the general trend of Clavius's proposals. Each student would study the basics of Euclid's *Elements*, and thereafter learn more advanced topics sporadically. In addition, "those apt and inclined to mathematics should be trained privately, after the course." Clavius, in the end, had not gotten his own school of mathematics, but he still got much of what he wanted.

Clavius's dogged advocacy of mathematics was never limited to the question of the curriculum; he also threw himself into the daunting project of writing new textbooks to replace the medieval texts in use in the Society's schools. While these were considered authoritative, they also dated back hundreds of years, and presented material in a style that was unlikely to appeal to sixteenth-century students. In 1570, Clavius published the first edition of his commentary on Sacrobosco's *Tractatus de Sphaera*, the standard medieval astronomy textbook, and in 1574 the first of many editions of his commentary on Euclid. These were followed by books on the theory and practice of the gnomon—the vertical part of a sundial—in 1581; the astrolabe—used for measuring the height of a star above the horizon—in 1581; practical geometry (1604); and algebra (1608). The textbooks were often clothed in the guise of commentaries on the traditional texts, such as Euclid's *Elements* and Sacrobosco's *de Sphaera*, and indeed they did retain the core teachings of their sources (such as Sacrobosco's assumption that the sun revolves around the Earth). Nevertheless Clavius's editions were in effect new books, bringing in new and up-to-date topics, emphasizing applications, and presenting the materials in a clear and appealing manner. They saw many editions throughout the sixteenth and seventeenth centuries, and remained the standard textbooks in the Jesuit schools well into the seventeen hundreds.

The project closest to Clavius's heart, however, was the establishment of a mathematics academy at the Collegio Romano. Initially, in the 1570s and '80s this was an informal group of select mathematics students who gathered around Clavius to study advanced topics. But in the early 1590s, Clavius managed to convince his friend the theologian Robert Bellarmine, who was the Collegio's rector at the time, to formalize the arrangement. Thereafter members of the academy were exempted from other duties for a year or two during their studies, and allowed to concentrate exclusively on mathematics. In 1593, General Acquaviva lent his own authority to the arrangement, decreeing that the best mathematics students in the Jesuit network of colleges would be sent to Rome to study with Father Clavius. The result was that Clavius was soon the leader of a group of young mathematicians who were not only competent teachers, but brilliant mathematicians in their own right. Among them was the statesmanlike father Christoph Grienberger, who was Clavius's successor at the Collegio; the fiery father Orazio Grassi (1583–1654), who famously tangled with Galileo over the nature of comets; Father Gregory St. Vincent (1584–1667); and Father Paul Guldin (1577–1643)—all of them among the foremost European mathematicians of their generation. In 1581, Clavius had complained that Jesuits were ignorant of mathematics and fell silent when it was discussed. But owing almost entirely to his dogged and tireless leadership, only a few decades later, Jesuits were setting the standard for the study of mathematics in Europe.

Through it all, even in their advanced work, the Jesuits never deviated from their commitment to Euclidean geometry. It was the core of their teaching and the foundation of their mathematical practice. This was not a stylistic choice, but a deeply held ideological commitment: the whole point of studying and teaching mathematics was that it demonstrated how universal truth imposed itself upon the world—rationally, hierarchically, and inescapably. Ideally, the Jesuits believed, the truths of religion would be imposed on the world just like geometrical theorems, leaving no room for avoidance or denial by Protestants or other heretics and leading to the inevitable triumph of the Church. For the Jesuits, mathematics must be studied according to the principles and procedures of Euclid, or it should not be studied at all. A mathematics that ran counter to these practices not only was useless to their purposes, but it would challenge their unconquerable faith that truth, handed down through the hierarchy of the universal Catholic Church, would inevitably prevail.

THE SLOW-WITTED BEAST

Christopher Clavius died in Rome on February 12, 1612, at the height of his power and prestige. The struggles of his early years were far behind him, and "our Clavius," as he was referred to in Jesuit documents, was one of the Society's cherished treasures. He was the undisputed founder and leader of the brilliant mathematical school, which not only brought honor to the Jesuits but also increased their political clout when they tried to establish themselves as the intellectual authority of the Catholic Church. Even their great rivals the Dominicans could boast of no comparable accomplishment. As recently as 1610, Clavius was called upon to confirm or deny Galileo's astounding telescopic observations, including his contention that there were mountains on the moon and that Jupiter was orbited by four moons. Clavius's intervention was decisive: he supported Galileo, thereby ensuring that the discoveries were accepted almost universally as fact.

The Jesuits' reverence for the aging Clavius comes through in the words of Jesuit astronomer Giambattista Riccioli, who commented in 1651 that "some would rather be blamed by Clavius than praised by others." His many admirers outside the Society included the Danish astronomer Tycho Brahe, the Italian mathematicians Federico Commandino and Guidobaldo del Monte, and as eminent a figure as the Archbishop of Cologne, who wrote in 1597 that Clavius is regarded as "the father of mathematics" and is "venerated by the Spanish, the French, the Italians, and most Germans." But Clavius did not lack for detractors either. Some, as would be expected, were Protestants, such as the German astronomer and mathematician Michael Maestlin, best known as Kepler's mentor, who was harshly critical of the calendric reform. So was the French humanist Joseph Justus Scaliger, who despised all Jesuits and referred to Clavius as "a German beast with a big belly, slow witted and patient."

Others, however, were Catholic. Cardinal Jacques Davy Duperron also found the livestock analogy useful, referring to Clavius as the "fat horse of Germany," and the French mathematician François Viète, who got into a fierce fight with Clavius over the merits of the new calendar, denounced him as "a false mathematician and a false theologian."

Such vitriolic denunciations, considered beyond the pale of academic discourse today, were not unusual in the sixteenth and seventeenth centuries. But even so, the references to Clavius as a slow-witted "beast" or "horse" were not idle insults for a man known for his girth. They bespoke a deeper criticism of Clavius, one that could not easily be brushed aside. Jacques-Auguste de Thou expresses it very clearly in his *History* of 1622, when he cites Viète's view of the Jesuit: Clavius, de Thou writes, was a master expositor who possessed a talent for explaining the discoveries of others, but made no original contributions to the disciplines over which he presided. He was, in this view, nothing more than a beast of burden, capable of immense expenditures of energy on behalf of his cause but incapable of original insight.

And this, it must be said, is not an entirely unjust assessment. Clavius was, unquestionably, a great promoter of the mathematical sciences, raising their profile both inside and outside the Society. He was an

CHRISTOPHORVS · CLAVIVS

"The fat horse of Germany." Christopher Clavius around 1606. Engraving by
E. de Boulonois, after a painting by Francisco Villamena. From I. Bullart,
Académie des Sciences (Amsterdam: Daniel Elzevier, 1682).
(Photograph courtesy of the Huntington Library)

effective organizer, who powered through political and organizational obstacles to establish his mathematical institute at the Collegio Romano. He was a master teacher, beloved and revered by generations of students, quite a few of whom became leading mathematicians in their own right. He was one of the leading pedagogues of the age, whose detailed mathematical curriculum profoundly shaped the teaching of mathematics in Europe for years to come. And perhaps most influentially, he was an author of textbooks, issuing repeated editions of his books on geometry, algebra, and astronomy.

But was he a creative mathematician? His textbooks offer little evidence of this. His *Euclid* is essentially a latter-day exposition of the ancient text, though it has been pointed out that it contains some new results in the theory of combinations. His edition of Sacrobosco's *Tractatus de Sphaera* does make use of some observations and theories that postdate the medieval original, but at a time when the traditional geocentric worldview was being challenged by Nicolaus Copernicus, Tycho Brahe, and Johannes Kepler, Clavius's text is a strict defender of the old orthodoxy. And although he knew of Viète's groundbreaking work, which is the foundation of modern algebra, Clavius's *Algebra* contains no trace of it, summarizing instead the ideas of earlier Italian and German algebraists whose work pales by comparison. All of which is to say that de Thou's depiction of Clavius as an unoriginal mathematician who never strayed far from the well-beaten path of his predecessors is supported by the evidence. And while this depiction was undoubtedly meant to insult the old Jesuit, it appears particularly severe in our own age, when mathematicians are judged almost exclusively by their creativity and originality.

To judge Clavius by this standard, however, would be an injustice. Clavius never wanted to make any original contributions to mathematics, and would have been quite happy if no one else did, either. "The theorems of Euclid and the rest of the mathematicians," he explained in the "Prolegomena," "still today as for many years past, retain . . . their true purity, their real certitude, and their strong and firm demonstrations." Mathematics is worth studying, for Clavius, not because it offers a field for open-ended investigations and new discoveries, as it does for a modern mathematician, but precisely because it never changes: its results are as true today as they were in the faraway past, and as they will be in the distant future. Mathematics, more than any other field, offers stability, order, and unchanging eternal truths. New discoveries are not only irrelevant for this goal, but potentially disruptive, and should by no means be encouraged. Judged from this standpoint, Clavius may indeed have been one of the great mathematicians of his age, but he was a very different kind of mathematician from those we know today.

In sticking close to the old and established truths of mathematics, Clavius was being true to the intellectual traditions of his order and the decrees of its leaders. No one, General Acquaviva had warned, should even suspect the Jesuits of innovating, and while this entrenched conservatism applied to all fields of knowledge, it was particularly critical in mathematics. The entire case that Clavius had made for raising the status of mathematics in Jesuit schools rested on the fact that mathematics, more so than any other science, was fixed, orderly, and eternally true. Other fields might be pursued for other reasons: theology, because it was the study of the word of God; philosophy, because it was the study of the world and was essential for understanding theology. But why study mathematics? Only because it provided a model of perfect rational order and certainty, and an example of how universal truths governed the world. If mathematics was to become a field of far-reaching innovation, in which new truths were proposed and then subjected to challenge and debate, then it would be worse than useless. It would be dangerous, as it would compromise the very foundations of the truth it was meant to buttress.

Clavius's imprint remained strong within the Jesuit mathematical tradition. For centuries, the Society's mathematicians chose to stick to tried-and-true methods, following Euclid as much as possible and avoiding treacherous new fields. But in the very years that Clavius was establishing the powerful Jesuit school of mathematics, a very different mathematical practice was gaining ground that would put all his cherished principles to the test. Where the Jesuits insisted on clear and simple postulates, the new mathematicians relied on a vague intuition of the inner structure of matter; whereas the Jesuits celebrated absolute certainty, the new mathematicians proposed a method rife with paradoxes, and seemed to revel in them; and whereas the Jesuits sought to avoid controversy at all cost, the new method was mired in intractable controversies seemingly from its very inception. It was everything that the Jesuits thought mathematics must never be, and yet it flourished, gaining new ground and new adherents. It was known as the method of indivisibles.

3

Mathematical Disorder

THE SCIENTIST AND THE CARDINAL

In December of the year 1621, Galileo Galilei, mathematician and philosopher at the court of Grand Duke Ferdinando II of Tuscany, received a letter from an admirer, the twenty-three-year-old Milanese monk Bonaventura Cavalieri. Galileo had met Cavalieri in Florence some months before, was impressed with the young monk's mathematical acumen, and invited him to continue their exchanges by post. Cavalieri did just that; his letter, filled with admiration and praise for the Florentine sage, reported on his latest mathematical work and asked for Galileo's opinion on the radical new direction he was taking.

Galileo at the time was at the height of his power and fame, and was used to ambitious young men seeking his advice and patronage. It had been twelve years since, while still a mathematics professor at the University of Padua, he had built a telescope and pointed it at the skies. What he saw there changed humans' view of the universe forever: innumerable stars invisible to the naked eye, mountains and valleys on the supposedly spherical disk of the moon, dark spots on the allegedly perfect surface of the sun. Most remarkable were the four tiny specks he observed circling Jupiter, which he inferred were moons orbiting the planet just as our moon orbits Earth. Galileo had quickly compiled his findings in a booklet he titled *Sidereus nuncius* (*The Starry Messenger*) and sent it to the leading scholars and astronomers of the day. The impact was immediate, and seemingly overnight the obscure professor became known across Europe as the man who had opened up the heavens. On visiting Rome in 1611, Galileo regaled the Pope with tales of his discoveries, and was invited to a friendly private audience with the Jesuit cardinal Bellarmine. At the Collegio Romano, Father Clavius, ever suspicious of innovation, initially demurred, quipping that in order to see such things one must first have put them inside the telescope. But he, too, came around when Jesuit astronomers confirmed the Florentine's discoveries and gave them their blessings. The venerable mathematician, now in the last year of his life, was in attendance when the Jesuits celebrated Galileo with a day of lavish ceremonies at the Collegio.

Like many a professor then and now, Galileo was not enamored of his university teaching duties. Sensing an opportunity to be rid of this burden once and for all, he dedicated the *Sidereus nuncius* to the ruler of his native Florence, the grand duke Cosimo II de'Medici of Tuscany, hinting broadly that he would like nothing better than to join that prince's household. To sweeten things further, he named the newly discovered satellites of Jupiter, in honor of the grand duke and his family, "the Medicean Stars," thus inscribing the Medici name in the heavens for all time. The gambit worked: by 1611, Galileo had left Padua for the Medici court in Florence, where he was appointed chief mathematician and philosopher to the grand duke. Happily for Galileo, his new position came with no teaching obligations, even though he was officially named the head mathematician of the University of Pisa. It also came with a salary several times greater than he had enjoyed as a lowly professor.

Now famous, Galileo did not rest on his laurels. A flamboyant man with a quick wit and a sharp pen, he enjoyed the rough-and-tumble of scientific disputes. Not long after moving to Florence, he wrote his *Discourse on Floating Bodies*, which was, in effect, a direct attack on the principles of Aristotelian physics. By 1613 he had published his *Letters on Sunspots*, which tells the tale of his debate with the mysteriously named "Apelles" over the discovery and nature of this solar phenomenon. In it, Galileo argues that he was the first to observe sunspots (a

Galileo at the height of his fame. Portrait by Ottavio Leoni (1578–1630).
(RMN-Grand Palais / Art Resource, NY)

claim that may have been sincere but was in any case mistaken). He also argues, correctly, that the spots are on the surface of the sun or very close to it, and that they demonstrate that the sun is rotating on its axis. Finally, moving beyond the immediate matter at hand, he claims the sunspots provide crucial support to the Copernican system, which placed the sun, and not the Earth, at the center of the cosmos. As Galileo knew well, these claims were sure to rile traditional Aristotelian scholars, who believed that the heavens were perfect and that the spots must be an atmospheric effect near the Earth. Things got even testier when "Apelles" was revealed as the Jesuit scholar Christoph Scheiner of Ingolstadt, who was deeply offended by Galileo's ridicule. It was the first sign of friction between Galileo and the Jesuits, and it came only two years after he was publicly honored at the Collegio Romano. But it was far from the last, as the tensions between Galileo and the Society of Jesus would only grow in the years that followed. When, nearly twenty years later, the Florentine scientist would be put on trial by the Inquisition, accused and ultimately convicted of heresy, it was the Jesuits who led the charge.

Galileo was treading on dangerous ground. He was not only challenging the authority of Aristotle, held dear by the Church theologians, but also going against the clear meaning of Scripture, which in several places implied that the sun revolves around the Earth. A more squeamish soul might have steered clear of such a potentially explosive issue, but Galileo was anything but. Instead of waiting for the attacks of his adversaries, he decided to take the battle onto their home ground by publishing his own theological treatise. The "Letter to the Grand Duchess Christina" was addressed to Christina of Lorraine, mother of the ruling grand duke of Tuscany, who had expressed her concerns to Galileo that his system was inconsistent with the revealed word of God. Circulated in 1615, but published only many years later, the letter contains Galileo's response, which became known as the doctrine of "the two books." The book of nature, he reasoned, and the book of Scripture can never be in conflict. One contains all we see around us in the world, the other contains divine revelation, but both are ultimately derived from the same source: God Himself. Therefore, if there appears to be conflict between the two, the only possible explanation is that we do not properly understand the one or the other.

As long as we do not have scientific "proof" of a particular thesis, Galileo conceded, we should always accept the authority of Scripture, understood in its most simple and direct meaning. But if we do possess a scientific proof, then the roles are reversed, and Scripture must be reinterpreted to accord with the book of nature. Otherwise, Galileo warned, we will be required to believe something that is manifestly false, bringing ridicule and discredit upon the Church. This, Galileo insisted, was precisely why the Church should accept Copernicanism. He could prove, he insisted, that Earth and the planets do indeed revolve around the sun, and the Church would only discredit itself by contradicting manifest truth. The traditional understanding of Scripture must be replaced with interpretations that are consistent with scientific truth, Galileo argued, and included his own readings of critical biblical passages to show that they were perfectly consistent with Copernicanism.

Beautifully written and eminently persuasive, the "Letter to the Grand Duchess Christina" is a cogent defense not just of Copernicanism, but of the compatibility of faith and free scientific research. But seventeenth-century religious authorities were not inclined to look kindly on an uninvited intrusion onto their turf. Granted, Galileo was a gifted astronomer, but he had no business pronouncing on theology, a field in which he was strictly an amateur. The task of reminding the interloper of his place fell to Clavius's old friend, the venerable Jesuit theologian cardinal Robert Bellarmine. In April of 1615, the cardinal issued an opinion on a work by one of Galileo's ardent followers, the Carmelite monk Paolo Foscarini. Although nominally addressed to Foscarini, the opinion was clearly intended to put Galileo on notice. If there were scientific proof of Copernicanism, Bellarmine conceded in his letter, then passages in Scripture should be reconsidered, since "we should rather have to say that we do not understand them than to say something is false which had been proven." But since no such proof "has been shown to me," he continued, one must stick to the manifest meaning of Scripture and the "common agreement of the holy fathers." All of these agreed that the sun revolves around the Earth.

Bellarmine undoubtedly had a point. Galileo could and did bring up many strong arguments in support of the Copernican system, but despite his brave proclamations, he could not actually prove it. His supposed "proof," based on the ebb and flow of the tides, was weak and, as some contemporaries pointed out, deeply flawed. In the absence of proof, Bellarmine's insistence that Scripture be taken at face value seems eminently reasonable. He furthermore did not prohibit Galileo from studying the Copernican system as a hypothesis that fit well with observations. He only insisted that Galileo not hold that Copernicanism was, in fact, true, and that it described the actual motions of the sun and planets.

Less than a year after Bellarmine wrote the letter, his opinion became the official Church position, putting strict limits on Galileo's ability to advocate for Copernicanism. But in 1616 the Church was not ready to give up on its erstwhile hero, who was not only the most celebrated scientist in Europe but also a good Catholic. In a sign of the high esteem in which he was held in Rome, Galileo was granted an interview with Pope Paul V, who assured him of his goodwill, and with Bellarmine, who explained the terms of the ban and confirmed them in writing. It was Galileo's alleged violation of this injunction that brought him before the Inquisition sixteen years later.

In the years that followed, Galileo had to all appearances put this distressing affair behind him. He was still a famous man, admired by scientists and laymen alike, and secure in his position in the Medici court. His skirmishes with Church authorities made him suspect in some circles, the Jesuits above all. But they also made him a hero of more liberal segments of Italian society, who resented the Church's insistence that it was the arbiter of all truth, and resented even more the domineering ways of the Jesuits. The bastion of this "liberal party" was the Accademia dei Lincei (Academy of the Lynx-Eyed) in Rome, of which Galileo was the most illustrious member. Founded in 1603 by the aristocratic Federico Cesi, the academy was a gathering place for some of the most brilliant intellectuals in Rome, both ecclesiastics and laymen. During the troubled years of 1615–16, the Linceans stood shoulder to shoulder with Galileo, and their support no doubt helped him get off as lightly as he did. They would prove just as important in the following years, as Galileo began once again to express forbidden views of the heavens.

PARADOXES AND INFINITESIMALS

In 1621, Galileo, still cautious, was likely pleased to receive an inquiry about a seemingly safe mathematical topic. Suppose, Cavalieri suggested, that we have a plane figure and we draw a straight line inside it, and suppose, furthermore, that we then draw all the possible lines inside the figure that are parallel to the first. "In that case," he writes, "I call the lines so drawn 'all the lines' of that plane figure. Similarly, given a three-dimensional solid, all the possible planes inside a solid that are parallel to a given plane are 'all the planes' of that solid." Is it permissible, he inquires of Galileo, to equate the plane figure with "all the lines" of the figure and the solid with "all the planes" of the solid? Furthermore, if there are two figures, is it permissible to compare "all the lines" of one with "all the lines" of the other, or "all the planes" of one with "all the planes" of the other?

Cavalieri's question seems simple, but it goes straight to the paradoxical heart of the infinitely small. On an intuitive level, the plane does indeed seem to be composed of parallel lines, and a solid appears to be composed of parallel planes. But as Cavalieri notes in his letter, we can draw an infinite number of parallel lines through any figure, and an infinite number of planes through any solid, which means that the number of "all the lines" or "all the planes" is always infinite. Now, if each of the lines has a positive width, however small, then an infinite number of them will add up to an infinitely large figure—not to the one we started out with. But if the lines have no width (or zero width), then any accumulation of them, no matter how large, will still have zero width and zero magnitude, and we are left with no figure at all. The same applies to "all the planes" of a three-dimensional solid: if they have a thickness, however small, they will inevitably combine to a solid of infinite size; but if they have no thickness, then any accumulation of them will always add up to zero.

It is the old question of the composition of the continuum that had confounded philosophers and mathematicians since the days of Pythagoras and Zeno. And to this familiar if troublesome one, Cavalieri now added another query: is it allowable to compare "all the lines" of one figure with "all the lines" of another? This, he notes in his letter, involves comparing one infinity with another, a move that was strictly forbidden by the traditional rules of mathematics. This is because, according to the "axiom of Archimedes," two magnitudes have a ratio if and only if one can multiply the smaller magnitude so many times that it will be bigger than the larger magnitude. This, however, is not the case with infinities, since, however many times one might multiply infinity, one will always arrive at the same unchanging result: infinity.

Unfortunately we do not have Galileo's response to his young colleague, since only one side of the correspondence has survived. Letters from Cavalieri in the following months suggest that Galileo, at the very least, encouraged Cavalieri to continue his investigations. This is likely what Cavalieri had expected, since Galileo was already reputed to hold unorthodox views on the composition of the continuum. As early as 1604, while working out the law of falling bodies, he had experimented with the notion that the surface area of a triangle, representing the distance traveled by a body, was composed of an infinite number of parallel lines, each representing the body's speed at a given instant. Some years later, in 1610, Galileo was still occupied with the paradoxes of the continuum, and announced his intention to devote an entire book to the matter. The book never materialized, probably because of the dramatic events that recast his life in those years, but three decades later he offered a fairly detailed exposition of his views in his last great work, *Discourses and Mathematical Demonstrations Relating to Two New Sciences*, considered by many today to contain his most important scientific contributions. The *Discourses*, as the work came to be known, was written in Galileo's villa in Arcetri, outside Florence, during the long years of house arrest that followed his condemnation by the Inquisition in 1633. Although published in Holland in 1638, it is based on studies that Galileo conducted many decades before, when he was a professor at the universities of Pisa and Padua.

The *Discourses* is written as a conversation between three friends, Salviati, Sagredo, and Simplicio, who would have been familiar to the book's readers. Only a few years before, the same trio had starred in the *Dialogue on the Two Chief World Systems*, Galileo's immensely popular book on the Copernican system and the work that had led to his trial and condemnation by the Inquisition. Although much of the sparkle and wit that characterized *Dialogue* is absent from the *Discourses*, the three friends retain their former roles: Salviati as Galileo's spokesman, Simplicio as the voice of Galileo's outdated Aristotelian critics, and Sagredo as the wise arbiter, who regularly sides with Salviati. In the first of the four days of the dialogue, the three friends discuss the question of cohesion: what is it that holds materials together, and prevents them from breaking up under outside pressures? Salviati begins by discussing ropes, showing that their strength is due to the fact that they are composed of a large number of threads packed and twisted together. He then extends his discussion to wood, whose inner strength, he argues, is also due to the fact that it is composed of tightly packed fibers. But what, he asks, of other materials, such as marble or metals? What force is it that holds them together with such remarkable strength?

The answer, according to Salviati, is "horror vacui"—nature's abhorrence of a vacuum. We know from experience, he argues, that *horror vacui* is an extremely powerful force: two perfectly smooth surfaces of marble or metal can hardly be separated, since pulling them apart would produce a momentary vacuum. This powerful force, he continues, is active not only between bodies, but also inside each body, holding it together. Just as a rope is composed of separate threads, and wood of separate fibers, a block of marble or a sheet of metal is also composed of innumerable atoms arranged side by side. There is, however, this difference: whereas a rope is composed of a large but finite number of threads, and a piece of wood of an even larger but still finite number of fibers, a block of marble or a sheet of metal is composed of an *infinite* number of infinitely small atoms, or "indivisibles." Separating them is an infinite number of infinitely small empty spaces. The vacuum in these infinite spaces is the glue that holds the object together and is responsible for its internal strength.

This was Salviati's (Galileo's) theory of matter, and as he himself admitted, it was a difficult one. "What a sea we are slipping into without knowing it!" Salviati exclaims at one point. "With vacua, and infinities, and indivisibles . . . shall we ever be able, even by means of a thousand discussions, to reach dry land?" Indeed, can a finite amount of material be composed of an infinite number of atoms and an infinite number of empty spaces? To prove his point that it could, he turned to mathematics.

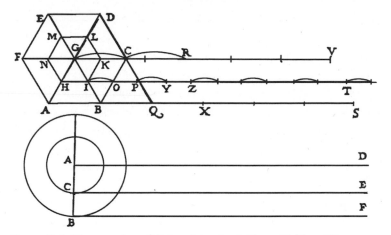

Figure 3.1. The paradox of Aristotle's wheel. From Galileo, *Discourses*,
Day 1 (*Le Opere di Galileo Galilei*, vol. 8, Edizione Nazionale
[Florence: G. Barbera, 1898], p. 68)

Salviati investigated the question of the continuum by way of a medieval paradox known as Aristotle's wheel, although it had nothing to do with that ancient philosopher, and its revelation was anything but Aristotelian. Imagine, Salviati suggests to his friends, a hexagon *ABCDEF* and a smaller hexagon *HIJKLM* within it and concentric with it, both around the center *G*. And suppose, furthermore, that we extend the side *AB* of the large hexagon to a straight line *AS*, and the parallel side of the smaller hexagon into the parallel line *HT*. Next we rotate the large hexagon around the point *B*, so that the side *BC* comes to rest on the segment *BQ* of the line *AS*. When this happens the smaller hexagon will also rotate until the side *IK* comes to rest on the segment *OP* of the line *HT*. There is a difference, Salviati points out, between the line created by the rotating large hexagon and the line created by the smaller hexagon: the larger hexagon is creating a continuous line, because the segment *BQ* is placed right next to the segment *AB*. The smaller hexagon's line, however, has gaps, because between the segments *HI* and *OP* there is a space *IO* where the hexagon in its rotation never touches the line. If we complete a full rotation of the large hexagon along the line *AS*, it will create a continuous segment whose length is equal to the hexagon's perimeter. At the same time, the smaller hexagon will travel a distance approximately equal in length along the line *HT*, but the line it creates will not be continuous: it will be composed of the six sides of the hexagon, with six equal gaps between them.

Now, what is true of hexagons, according to Salviati, is true of any polygon, even one with 100,000 sides. Rolling it along will create a straight line equal in length to its circumference, whereas a smaller but similar polygon inside it will trace a line of equal length, but composed of 100,000 segments interspersed with 100,000 empty spaces. But what will happen if we replace those finite polygons with a polygon with an infinite number of sides—in other words, a circle? As the lower part of Aristotle's wheel shows, rolling the circle one full revolution will trace a line *BF* equal to the circle's circumference, and the inner circle, meanwhile, will trace a line of equal length along *CE* while completing its own revolution. In this the circles are no different from the polygons. But here's the problem: the length of the line *CE* is equal to that of *BF*, a line created by a circle with a greater circumference. How can the smaller circle create a line longer than its own circumference? The answer, according to Salviati, is that the seemingly continuous line *CE* is, just like the line created by the rotating polygons, interspersed with empty spaces that contribute to its length. The line created by the smaller polygon of 100,000 sides is composed of 100,000 segments separated by 100,000 gaps; it follows that the line traced by the smaller circle is composed of an infinite number of segments separated by an infinite number of empty spaces.

By pushing Aristotle's wheel to its illogical limit, Galileo arrived at a radical and paradoxical conclusion: a continuous line is composed of an infinite number of indivisible points separated by an infinite number of minuscule empty spaces. This supported both his theory of the structure of matter and his view that material objects are held together by the vacuum that pervades them. It provided a new way of thinking about the material world and also pointed to a new vision of mathematics, one in which, according to Salviati, any "continuous quantity is built up of absolutely indivisible atoms." The inner structure of the mathematical continuum is indistinguishable from the threads of rope, the fibers inside blocks of wood, or the atoms that make up a smooth surface: it is composed of tightly compressed indivisibles with empty spaces between them. For Galileo, the mathematical continuum was modeled on physical reality.

Galileo's approach was troubling to contemporary mathematicians, as it went directly against the well-established paradoxes that had guided the treatment of the continuum since antiquity. He did, nonetheless, have at least one prominent supporter in fellow Lincean Luca Valerio (1553–1618), who had been inducted into the Accademia dei Lincei at Galileo's urging. Valerio was professor of rhetoric and philosophy at the Sapienza University in Rome, and widely acknowledged as one of the leading mathematicians in Italy. In his *De centro gravitatis* of 1603 and *Quadratura parabola* of 1606 he had experimented extensively with indivisibles, which enabled him to determine the centers of gravity for plane figures and solids.

But Valerio learned his mathematics among the Jesuits of the Collegio Romano, under the tutelage of Clavius himself, and with this prominent group, Galileo's mathematical atomism found no favor. For the Jesuits, indivisibles represented the exact inverse of the proper and correct approach to mathematics. The Jesuits, it will be recalled, valued mathematics for the strict rational order it imposed upon a seemingly unruly universe. Mathematics, and in particular Euclidean geometry, represented the triumph of mind over matter and reason over the untamed material world, and reflected the Jesuit ideal not only in mathematics but also in religious and even political matters. By anchoring his mathematical speculations in an intuition of the structure of matter rather than in self-evident Euclidean postulates, Galileo turned this order on its head. The composition of the mathematical continuum, according to Galileo, could be derived from the composition of ropes and the inner structure of a piece of wood, and it could be interrogated by imagining a wheel rolling over a straight surface. In place of the Jesuit approach, Galileo proposed that geometrical objects such as planes and solids were little different from the material objects we see around us. Instead of mathematical reason imposing order on the physical world, we have pure mathematical objects created in the image of physical ones, incorporating all their incoherence. Clavius, needless to say, would not have been pleased.

Although Galileo's support of infinitesimals gave them a degree of visibility and respectability that no other endorsement could have achieved, he himself made little use of indivisibles in his actual mathematical work. One of the few exceptions is in his famous argument on the distance traversed by a free-falling body. Suppose, Salviati proposes in Day 3 of *Discourses*, a body is placed at rest at point *C*, and then accelerates at a constant pace, as in free fall, until it reaches point *D*. Now let the line *AB* represent the total time it takes that body to get from *C* to *D*, and the line *BE*, perpendicular to *AB*, represent the body's greatest speed, which it reaches at *D*. Draw a line from *A* to *E*, and lines parallel to *BE* at regular intervals between *AB* and *AE*. Each of these lines, Salviati argues, represents the speed of the object at a particular moment during its steady acceleration. Since there is an infinite number of points on *AB*, each representing an instant in time, there is also an infinite number of such parallel lines, which together fill in the triangle *ABE*. The sum of all the speeds at every point, furthermore, is equivalent to the total distance traversed by the object during the time *AB*.

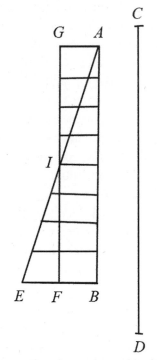

Figure 3.2. Galileo on uniformly accelerated bodies. From *Discourses*, Day 3. (Ed. Naz., vol. 8, p. 208)

Now, Salviati says, if we take a point *F* halfway between *B* and *E*, and draw a line through it parallel to *AB*, and a line *AG* parallel to *BF* intersecting with it, then the rectangle *ABFG* is equal in area to the triangle *ABE*. But just as the area of the triangle represents the distance traversed by a body moving at a uniformly accelerated speed, so the area of the rectangle represents the distance covered by a body moving at a fixed speed. It follows, Salviati concludes, that the distance covered in a given time by a body that begins at rest and uniformly accelerates is equal to the distance covered by a body moving at a fixed speed for the same amount of time, if the speed is half the maximum speed reached by the accelerated body.

Known as the law of falling bodies, it is one of the first things any student learns today in a secondary school physics class, but in its time, it was nothing short of revolutionary. It was the first quantitative mathematical description of motion in modern science, and it laid the foundations for the modern field of mechanics—and, in effect, modern physics. Galileo was well aware of the importance of the law, and he included it in two of his most popular works, the *Dialogue* of 1632 and the *Discourses* of 1638. Although it relies mostly on Euclidean geometrical relations, it does show Galileo's willingness to assume that a line is composed of an infinite number of points. That was precisely the question posed to him by Cavalieri in 1621, and whatever answer Galileo gave him, the young monk was not discouraged. During the 1620s he took the idea of the infinitely small and turned it into a powerful mathematical tool that he called the method of indivisibles. The name stuck.

THE DUTIFUL MONK

Cavalieri was born in Milan in 1598 to a family that was likely respectable and possibly even noble, but of modest means. His parents named him Francesco, but he took the name Bonaventura at the age of fifteen, when he became a novice in the order of the Apostolic Clerics of St. Jerome, more commonly known as the Jesuats. Only a single letter distinguishes the Jesuats from Ignatius's famed Jesuits, but the two societies could hardly have been more different. Whereas the Jesuits were a modern order, forged in the crucible of the Reformation crisis, the Jesuats dated back to the fourteenth century and were the product of the fierce piety of the decades that followed the Black Death. Whereas the Jesuits were a dynamic force whose schools and missions encompassed the globe, the Jesuats were a local Italian order, respected for their work with the sick and dying but wholly lacking the ambition of Ignatius's followers. The forming of a Jesuit, as we have seen, could take decades, but the training of a Jesuat was a much briefer affair: in 1615, at the age of seventeen and two years into his novitiate, Cavalieri pronounced his vows and donned the white habit and a dark leather belt that identified him as a full-fledged member of the order. A few months later he left his home city of Milan for the Jesuat house in Pisa.

We do not know if the move to Pisa was Cavalieri's idea or that of his superiors, but it would turn out to be an auspicious one for the young Jesuat and for mathematics. "I am proud, and will always be," he wrote many years later to fellow mathematician Evangelista Torricelli (1608–47), "of having received under the serenity of that sky the first aliments and elements of mathematics." The instigator of his budding fascination with mathematics was Benedetto Castelli (1578–1643), Galileo's former student and lifelong friend and supporter, who was at the time the professor of geometry at the University of Pisa. Castelli introduced Cavalieri both to Galileo's work in physics and mathematics and, in due course, to the great Florentine himself. In 1617, Cavalieri moved to Florence, where, aided by the influence of his Milanese patron, Cardinal Federico Borromeo, he joined the circle of disciples and admirers around Galileo at the Medici court. "With your help," the cardinal wrote to Galileo, Cavalieri "will reach that level in his profession which we can already perceive from his singular inclinations and ability."

The following year, Cavalieri returned to Pisa, where he began giving private lectures in mathematics, substituting for Castelli, who was drafted into Grand Duke Cosimo's household to serve as tutor to his sons. Cavalieri was now a professional mathematician in all but title, but for the next decade his life was torn between his field of choice and his duties to the Jesuat order. In 1619 he applied for the mathematics chair at the University of Bologna, which had been vacant since Giovanni Antonio Magini died two years before. Only the active support of Galileo could have secured such a prestigious position for so young an applicant, but Galileo seemed reluctant to intervene, so the opportunity slipped away. Instead, in 1620, Cavalieri was recalled to the Jesuat house in Milan, where he also became deacon to Cardinal Borromeo. Far from the brilliant Medici court, Cavalieri found that his talents were not always appreciated. "I am now in my own country," he wrote to Galileo, "where there are these old men who expected of me great progress in theology, as well as in preaching. You can imagine how unwillingly they see me so fond of mathematics."

Despite his growing immersion in mathematics, Cavalieri was serious about his religious vocation. He set out to study theology, and soon made up for lost time, "to the great wonder of everyone." As a result, and also thanks to the cardinal's support, he rose quickly in the ranks of the order, and in 1623 he was named prior of the Jesuat monastery of St. Peter in the town of Lodi, not far from Milan. Three years later he was promoted to prior of St. Benedict's monastery in the larger city of Parma. Yet all the while, Cavalieri was casting about for a position as a professional mathematician. In 1623 he renewed his efforts to secure the professorship in Bologna, but the Bolognese Senate, while not rejecting his appeal outright, repeatedly asked him for more and more samples of his work. When his old mentor Castelli was appointed to the mathematics chair at Sapienza University in 1626, Cavalieri sensed an opportunity. But despite his taking leave of his duties to promote his case, and spending six months in Rome with Galileo's friend and fellow Lincean, the influential Giovanni Ciampoli (1589–1643), nothing came of it. Back in Parma, he approached the Jesuit fathers who ran the University of Parma, but as he wrote to Galileo after the fact, they would not allow a mere Jesuat, not to mention a student of Galileo, to teach in the university.

It was not until 1629 that the tide finally turned for Cavalieri. Galileo, at long last warming to his student's cause, declared that "few scholars since Archimedes, and perhaps nobody, have gone so deeply and profoundly into the understanding of geometry" as Cavalieri. The Bolognese Senate was duly impressed, and on August 25 it offered the vacant chair of mathematics at the University of Bologna to the Jesuat. Having spent a full decade trying to secure the position, Cavalieri did not hesitate: he quickly moved into the Jesuat house in Bologna, and began lecturing at the university that same October. He would stay in that city for the remaining nineteen years of his life, living in the monastery and teaching at the university. Although a young man by modern standards, he was in failing health and suffering from repeated bouts of gout, which made travel extremely difficult. Only once during those years did he venture from his adopted city, and it was for the only cause that could have lured him from the comfort of his daily routine: it was in 1636, when he visited Galileo during the old master's long and lonely years of house arrest.

The decade between Cavalieri's stay in Pisa and his appointment as professor in Bologna was an uncomfortable one for the young monk, but it was also his most mathematically productive period. In fact, nearly all the original proofs for which he became known, and even much of the actual text of his books, date to these itinerant years. Once settled in Bologna, he was weighed down by his teaching duties, as well as by the demands of the senate, which required its professor of mathematics to produce a steady stream of astronomical and astrological tables. Even so, the industrious monk managed to publish *Lo specchio ustorio* ("The Burning Mirror") in 1632, *Geometria indivisibilibus* ("Geometry by Indivisibles") in 1635, and *Exercitationes geometricae sex* ("Six Geometric Exercises") in 1647. These works, conceived and largely written during the 1620s, established Cavalieri's reputation as a mathematician, and the leading proponent of infinitesimals.

ON THREADS AND BOOKS

Just as Galileo began his mathematical theorizing on the continuum with a discussion on the inner composition of ropes and blocks of wood, Cavalieri, too, founded his mathematical method on our material intuitions: "It is manifest," he writes, "that plane figures should be conceived by us like cloths woven of parallel threads; and solids like books, composed of parallel pages." Any surface, no matter how smooth, is in fact made up of minuscule parallel lines, arrayed side by side; and any three-dimensional figure, no matter how solid it appears, is nothing but a stack of razor-thin planes, one on top of the other. These thinnest of

slices, equivalent to the smallest components, or atoms, of material figures, Cavalieri called indivisibles.

As he was quick to point out, there are important differences between physical objects and their mathematical cousins: a cloth and a
book, he noted, are composed of a finite number of threads and pages, but
planes and solids are made up of an indefinite number of indivisibles. It
is a simple distinction that lies at the heart of the paradoxes of the continuum, and whereas Galileo glossed over the matter in the *Discourses*,
the more cautious Cavalieri brought it to the fore. Even so, it is clear that
Cavalieri, like Galileo, began his mathematical speculations not with
abstract universal axioms, but with lowly matter. From there he moved
upward, generalizing our intuitions of the material world and turning
them into a general mathematical method.

For a taste of Cavalieri's method, consider proposition 19 in the first
exercise of the *Exercitationes*:

> If in a parallelogram a diagonal is drawn, the parallelogram is double
> each of the triangles constituted by the diagonal.

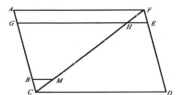

Figure 3.3. Cavalieri, *Exercitationes*, p. 35, prop. 19. (Bologna: Iacob Monti, 1647)

This means that if a diagonal *FC* is drawn for the parallelogram *AFDC*,
the area of the parallelogram is double the area of each of the triangles
FAC and *CDF*. If one approaches the proof in a traditional Euclidean
manner, then it is almost trivial: the triangles *FAC* and *CDF* are congruent, because, first, they share the side *CF*; second, the angle *ACF* is
equal to the angle *CFD* (because *AC* is parallel to *FC*); and third, the
angle *AFC* is equal to *DCF* (because *AF* is parallel to *CD*). Since the two
triangles together compose the parallelogram, and since, being congruent, they are equal in area, it follows that the area of the parallelogram is
double that of each of them. *QED*.

Cavalieri, of course, knew all this very well, and he likely would not have wasted a theorem in his book on proving something so elementary. But he was after something else, so he proceeded differently:

Let equal segments FE and CB be marked off from points F and C along the sides FD and CA respectively. And from the points E and B mark segments EH and BM, parallel CD, which cross the diagonal FC at points H and M respectively.

Cavalieri then shows that the small triangles *FEH* and *CBM* are congruent, because the sides *BC* and *FE* are equal, angle *BCM* is equal to *EFH*, and angle *MBC* is equal to *FEH*. It follows that the lines *EH* and *BM* are equal.

In the same way we show of the other parallels to CD, namely those that are marked in equal distances from the point F and C along the sides FD and AC, that they are also equal between themselves, just as the extremes, AF and CD, are equal. Therefore all the lines of the triangle CAF are equal to all the lines of the triangle FDC.

Since "all the lines" of one triangle are equal to "all the lines" of the other, Cavalieri argues, their areas are equal, and the parallelogram is double the area of each of them. *QED*.

The contrast between Cavalieri's proof and the traditional Euclidean demonstration is stark. The Euclidean proof began with the universal characteristics of a parallelogram, which are themselves derived from Euclid's self-evident postulates. From this universal beginning it moved step by logical step to establish the relations in this particular case—that of a parallelogram divided into two triangles. It shows, in essence, that the universal laws of reasoning *require* that the two triangles be equal. But Cavalieri refuses to proceed from such abstract universal principles, and begins instead with a material intuition: what, he asks, is the area of each triangle made of? His answer, based on a rough analogy to a piece of cloth, is that it is composed of parallel lines laid out neatly side by side. To find out the total area of each triangle, he then proceeds to "count" the lines that make it up. Since there is an infinity of lines in each surface, literally counting is impossible, but

Cavalieri shows that their number and size are nevertheless the same from one triangle to the other, and hence the areas of the two triangles are equal.

The point of Cavalieri's proof is to show not that the theorem is true—which is obvious—but rather *why* it is true: the two triangles are equal because they are composed of the same number of identical indivisible lines placed side by side. And it is precisely this material take on geometrical figures that distinguishes Cavalieri's approach from the classical Euclidean one. The Euclidean approach orders geometrical objects, and ultimately the world, through its universal first principles and its logical method. Cavalieri's approach, in contrast, begins with an intuition of the world as we find it, and then proceeds to broader and more abstract mathematical generalizations. It can rightly be called "bottom-up" mathematics.

Cavalieri's parallelogram proof showed that his method of indivisibles worked, but not that there was any advantage to adopting it. Quite the contrary: he offered a long and convoluted demonstration of a theorem that could be proved in one or two lines using the traditional Euclidean approach. If all Cavalieri's proofs went to such great lengths to accomplish so little, it is unlikely that he would have found many followers to adopt his approach. But this of course was not the case: the parallelogram proof demonstrated the reliability of indivisibles. To demonstrate their power, Cavalieri turned to more difficult challenges.

The "Archimedian spiral," known since antiquity, is produced by a point traveling at a fixed speed along a straight line, while the line itself rotates at a fixed angular speed around the point of origin. In the diagram, the curve is traced by a point traveling steadily from A to E, while the line AE itself is rotating at a fixed rate around the central point A. After a single revolution the spiral arrives at point E, and encloses a "snail-shaped" area AIE inside the larger circle MSE, whose radius is AE. Cavalieri set out to prove that the area enclosed within the spiral AIE is one-third the area of the circle MSE. Archimedes had used his own ingenious approach to demonstrate that this was so. Cavalieri, however, approached the problem in a novel, intuitive manner, using indivisibles to transform the complex spiral into the familiar and well-understood parabola.

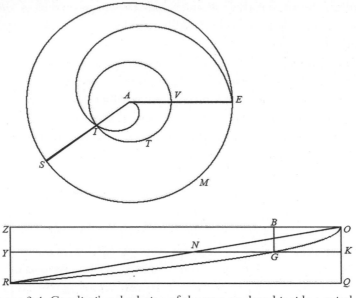

Figure 3.4. Cavalieri's calculation of the area enclosed inside a spiral.
From Cavalieri, *Geometria indivisibilibus* libri VI, prop. 19.
(Bologna: Clementis Ferroni, 1635)

Cavalieri posits a rectangle, *OQRZ*, in which the side *OQ* is equal to
the radius *AE* of the circle *MSE*, and the side *QR* is equal to the circle's
circumference. Returning to the spiral, he then picks a random point *V*
along *AE* and generates a circle *IVT* around the central point *A*. The circle
IVT has two parts: one, *VTI*, is outside the area enclosed by the spiral; the
other, *IV*, is inside the spiral. He takes the length *VTI* (external to the
spiral) and places it as a straight line *KG* inside the rectangle and parallel to *QR*, with *K* a point on the line *OQ*, and *OK* (that is, the distance
of *K* from *O*) equal to the radius *AV*. He then does the same for every
point along *AE*, taking the portion of its circle that is outside the spiral
and placing its length in its proper place along the side *OQ*, inside the
rectangle. Each point along *AE* has an equivalent point along *OQ*, with
a straight line emanating from it representing that portion of the circle
that lies outside the spiral. In the end, all the circular lines forming the
area *AES* outside the spiral are equal to all the straight lines composing
the area *OGRQ* inside the rectangle. Consequently, according to Cava-

lieri, the area enclosed by *OGRQ* is equal to the area inside the circle *MSE* that is not contained in the spiral.

What remains is to determine the area of the figure *OGRQ* (equal to the part of the circle outside the spiral) and compare it to the area of the entire circle. Cavalieri does this in two stages: First, using classical geometrical methods, he shows that the curve *OGR* is a parabola. Then, using indivisibles, he shows that the area of the triangle *ORQ* is equal to the area of the entire circle. This is clear if we consider the area of the circle to be made up of the circumferences of successive concentric circles, starting at the center (radius "0"), and culminating at the rim (radius *AE*). Placing the lengths of all these circumferences side by side, Cavalieri argued, produces the triangle *ORQ*. Now, he had previously shown that the area defined by a half parabola (*OGRQ*) is two-thirds the area of the enclosing triangle *ORQ*. Since *ORQ* is equal to the area of the entire circle, and *OGRQ* is equal to the area of the circle that lies *outside* the spiral, it follows that the area of the circle *inside* the spiral takes up the remaining area of the circle, or one-third of it. *QED*.

Cavalieri's proof of the area enclosed inside a spiral showed that his method could deal with the areas and volumes of geometrical figures, issues that were at the forefront of mathematical research at the time. Indeed, it demonstrated that indivisibles went to the very heart of geometrical questions, in a way that Euclidean proofs could not: indivisibles not only proved that certain relations held true, but also showed *why* this was the case. The two triangles that make up a parallelogram are equal because they are made up of the same indivisible lines; an Archimedean spiral encompasses one-third of its enclosing circle because its indivisible curves can be rearranged into a parabola. Whereas Euclidean proofs deduced necessary truths about geometrical figures, indivisibles allowed mathematicians to peer into the inner sanctum of geometrical figures and observe their hidden structure.

THE CAUTIOUS INDIVISIBLIST

Yet radical though his method was, Cavalieri was, by temperament and conviction, a conservative and quite orthodox mathematician. Deeply conscious of the logical conundrums presented by infinitesimals, he tried to burnish his orthodox credentials by staying as close as possible to the traditional Euclidean style of presentation. He also incorporated certain unwieldy restrictions into his method in an attempt to circumvent the paradoxes.

The internal tension within Cavalieri's work comes through in a letter he wrote to the aging Galileo in June of 1639. He had recently received a copy of Galileo's *Discourses*, and he was writing to thank the old master for his bold endorsement of indivisibles. Quoting the Roman poet Horace, Cavalieri compared Galileo to "the first to dare to steer the immensity of the sea, and plunge into the ocean," and then continued:

> It can be said that with the escort of good geometry and thanks to the spirit of your supreme genius, you have managed to easily navigate the immense ocean of indivisibles, of vacuum, of light, and of a thousand other hard and distant things that could shipwreck anyone, even the greatest spirit. Oh how much the world is in your debt for having paved the road to things so new and so delicate! . . . and as for me, I will not be a little obliged to you, since the indivisibles of my *Geometry* will gain indivisible lustre from the nobility and clarity of your indivisibles.

So far so good—Cavalieri is showering his old master with praise, and basking in the glow of his approval. Then, without warning, he takes a step back and renounces the very doctrine for which he has just praised Galileo. "I did not dare to affirm that the continuum is composed of indivisibles," he writes. All he did, he insists, was to show "that between the continua there is the same proportion as between the collection of indivisibles."

Cavalieri here comes remarkably close to disavowing his own indivisibles. Whereas in his books he boldly compared a geometrical plane to a piece of cloth woven with threads, and a solid to a book composed of pages, he now implies that he didn't really mean it. He took no position, he suggests, on the true composition of the mathematical continuum. All he did was introduce a new entity called "all the lines" of a plane figure or "all the planes" of a solid. Now, if a proportion exists between "all the lines" of one figure and "all the lines" of another, then, he claims, the same proportion exists between the areas of the two figures. And the same is true of "all the planes" of solids.

Cavalieri, assailed by critics, was insisting that he was agnostic on the thorny question of the composition of the continuum. His method, he insisted, was legitimate regardless of whether the continuous magnitudes were composed of indivisibles. He even avoided using the offending term itself. Remarkably, despite the fact that his most famous work is called *Geometry by Way of Indivisibles* (*Geometria indivisibilibus*), and although he discusses indivisibles in the methodological and philosophical passages of his works, he never actually mentions the term in his mathematical demonstrations, where the concept is always rendered as "all the lines" or planes. He placed strict limitations on the kinds of indivisibles that were allowed, and went out of his way to make his work appear traditional and orthodox, by presenting it in a traditional Euclidean mode of postulates, demonstrations, and corollaries. As for new and previously unknown results, Cavalieri avoided them altogether.

All, however, was to no avail. Cavalieri's contemporaries, whether hostile or sympathetic, simply did not believe his claim that he was undecided on the question of the composition of the continuum. His method, they thought, spoke for itself, and it clearly depended on the notion that continuous magnitudes are made up of infinitesimal components. Why would we be interested in a magnitude called "all the lines" if we didn't implicitly assume that these lines comprised a surface? Why would we compare "all the planes" of one solid with "all the planes" of another if we didn't think that they constituted their respective volumes? Cavalieri's bold metaphors of the cloth and the book, which openly endorse indivisibles, they found creative and inspiring, leading to ever-new discoveries. The cautious disclaimers that followed led only to an unwieldy terminology and a cumbersome method that largely negated the power and promise of indivisibles.

In the coming years, mathematicians who disliked Cavalieri's method, as did the Jesuits Paul Guldin and André Tacquet, denounced him for his violation of the traditional canons; those who welcomed his approach, as the Italian Evangelista Torricelli and the Englishman John Wallis did, claimed to be his followers while freely making use of infinitesimals with complete disregard for the Jesuat's carefully thought-out constraints. No one, but truly no one, actually followed Cavalieri's restrictive system.

Cavalieri's name and his books were often cited by mathematicians when they came under attack by critics of infinitesimals. The heavy and unwieldy volumes, with their contorted Latin, Euclidean structure, and air of solemn authority, provided some cover to later adherents of infinitesimal methods. They thought it was safe to point to the Jesuat master as the source of their system, and the one who had resolved all its difficulties in his learned volumes. After all, as they knew well, hardly anyone actually read Cavalieri's books.

GALILEO'S LAST DISCIPLE

In the end it was Cavalieri's younger contemporary, the brilliant Evangelista Torricelli, who took infinitesimals where the Jesuat would not go. Born in 1608 to a family of modest means, most likely in the city of Faenza, in northern Italy, young Evangelista moved to Rome at the age of sixteen or seventeen and there fell in love with mathematics. As he wrote in 1632 to Galileo, he did not receive a formal mathematical education but "studied alone, under the direction of the Jesuit fathers." Yet it was the Benedictine monk Benedetto Castelli—the same who had encouraged Cavalieri in his mathematical studies in Pisa—who was most influential in the young man's choice of vocation. Unlike his teacher Galileo, Castelli seemed to enjoy mentoring, and kept an eye out for promising young mathematicians. Now a professor at the Sapienza University in Rome, he took Torricelli under his wing and introduced him to the work of Galileo and Cavalieri.

In September 1632, no doubt with Castelli's encouragement, Torricelli wrote to Galileo, introducing himself as "a mathematician by profession, though still young, a student of Father Castelli for the past six years." The *Dialogue on the Two Chief World Systems* had appeared only a few months before, and the series of events that would lead to Galileo's condemnation and house arrest the following year was already under way. Torricelli begins by assuring the old master that Castelli takes every opportunity to defend the *Dialogue*, in order to avoid an "inconsiderate decision." He then moves to establish his own credentials as a geometer and astronomer, and a dedicated follower of Galileo's. "I was the first in Rome," he writes, to have studied your book assiduously and in detail . . . I did so with the pleasure that you can imagine for one who, already having a good enough experience of the geometry of Apollonius, of Archimedes, of Theodosius, and having studied Ptolemy and seen almost all of Tycho, Kepler, and Longomontanus, I adhered finally to Copernicus . . . and professed my attachment to the Galilean school.

Unfortunately for Torricelli, "ardent Galilean" proved to be a precarious identity in Rome once the *Dialogue* and its author were condemned less than a year later. This likely explains why we hear nothing of Torricelli for nearly a decade thereafter. He remained in Rome, pursued his mathematical work in private, studied Galileo's *Discourses on Two New Sciences*, which appeared in 1638, and generally kept a low profile. He reappears only in March 1641, when Castelli obtained permission to visit Arcetri, and wrote to Galileo to announce the good news. He will bring with him, he promised, a manuscript by the young Torricelli, who had been his student ten years before. "You will see," he flatters the old man, "how the road you have opened to the human spirit is followed by a very virtuous man. He shows us how fruitful and rich is the grain you have sown in this subject of motion; you will also see that he brings honor to the school of your Excellency."

The visions of open roads and fields of grain likely appealed to the lonely old man, who had been confined to his house for the past eight years. But it was the brilliance of Torricelli's work that had the greatest effect on Galileo. He was deeply impressed with what Castelli had shown him, and asked to meet the young mathematician. Castelli, for his part, was moved by Galileo's frailty and near blindness, and was concerned that he may not have long to live. Together they hatched a plan to bring Torricelli to Arcetri to serve as Galileo's secretary and help him edit and publish his latest works. Having received the invitation in early April, Torricelli wrote back to say that he was overcome and "confused" by the great honor done to him. Nevertheless, he seemed in no hurry to leave bustling Rome and join the old master in his lonely retreat. He made repeated excuses, but finally, in the autumn of 1641, he packed up his belongings and traveled to Galileo's Arcetri villa. There he spent his days editing the "fifth day" of the *Discourses*, to be added to the four days of dialogue that were published in 1638.

Only three months after Torricelli's arrival, his mission came to an abrupt end. In the early days of 1642, Galileo came down with heart palpitations and a fever, and on January 8, at the age of seventy-seven, the old master breathed his last. As a man condemned for "vehement heresy," he was interred in a small side room of the Basilica of Santa Croce in Florence, only to be moved to a place of honor in the central basilica a century later. Torricelli, meanwhile, was packing his things once more for the return journey to Rome when he received a startling offer: he could stay in Florence as Galileo's successor, and become mathematician to the Grand Duke of Tuscany and professor of mathematics at the University of Pisa. The offer did not include Galileo's position as court "philosopher," most likely because it was Galileo's insistence on his right, as philosopher, to pronounce on the structure of the world that had gotten him into trouble with the Church. But even without this additional accolade, the offer presented Torricelli with the opportunity of a lifetime: a secure position with a generous salary, the chance to pursue his studies without interruption, and public recognition as heir to the greatest scientist in Europe. He accepted without hesitation.

The next six years were remarkably productive for Torricelli. Previously he was so little known that Galileo had hardly heard of him, and Castelli had had to present him to Galileo as a former student. But with Galileo's passing and his appointment as mathematician to the Medici court, Torricelli suddenly became one of the leading scientists in Europe. He began a long and fruitful correspondence with French scientists and mathematicians, including Marin Mersenne (1588–1648) and Gilles Personne de Roberval (1602–75), and forged connections with fellow Italian Galileans Raffaello Magiotti (1597–1656), Antonio Nardi (died ca. 1656), and Cavalieri. Inspired by the *Discourses*, he pondered Galileo's thesis that nature's *horror vacui* (abhorrence of a vacuum) is what holds objects together. This led him in 1643 to experiments that established that a vacuum could, in fact, exist in nature, and to the construction of the world's first barometer.

Unlike the work of Galileo and Cavalieri, who published frequently, Torricelli's work can be found mostly in his correspondence and in unpublished manuscripts he circulated among his friends and colleagues. The sole exception is a book entitled *Opera geometrica*, published in

1644 and containing a collection of essays on subjects ranging from the physics of motion to the area enclosed by a parabola. Some of these, such as Torricelli's discussion of spheroids, rely on traditional mathematical methods derived from the ancients. The third treatise, however, entitled "De dimensione parabolae" ("On the Dimension of the Parabola"), is anything but traditional: it is Torricelli's dramatic introduction of his own method of indivisibles.

TWENTY-ONE PROOFS

Surprisingly, given its name, the purpose of "De dimensione parabolae" is not the calculation of the area inside a parabola. This was calculated and demonstrated by Archimedes more than 1,800 years before, and was well known to Torricelli and his contemporaries. It requires no further proof. What the treatise does offer is no fewer than twenty-one different proofs of this familiar result. Twenty-one times in succession, Torricelli poses the theorem that "the area of a parabola is four thirds the area of a triangle with the same base and height," and twenty-one times he proves it, each time differently. It is likely the only text in the history of mathematics to offer this many different proofs of a single result, and by a wide margin. It is a testament to Torricelli's virtuosity as a mathematician, but its purpose was different: to contrast the traditional classical methods of proof with the new proofs by indivisibles, thereby showing the manifest superiority of the new method.

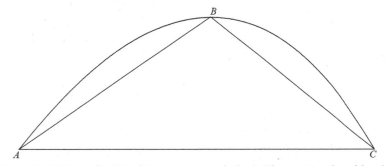

Figure 3.5. Torricelli, "De dimensione parabolae": The area enclosed by the parabola *ABC* is four-thirds the area of the triangle *ABC*.

The first eleven proofs of "De dimensione" conform to the highest standards of Euclidean rigor. To calculate the area enclosed in a parabola, they make use of the classical "method of exhaustion," attributed to the Greek mathematician Eudoxus of Cnidus, who lived in the fourth century BCE. In this method the curve of the parabola (or a different curve) is surrounded by a circumscribed and a circumscribing polygon. The areas of the two polygons are easy to calculate, and the area enclosed by the parabola lies somewhere in between. As one increases the number of sides of the two polygons, the difference between them becomes smaller and smaller, limiting the possible range of the area of the parabola.

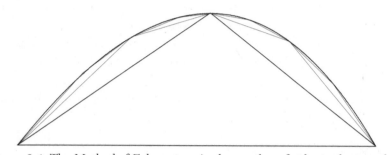

Figure 3.6. The Method of Exhaustion. As the number of sides in the inscribed polygon is increased, its area more closely approximates the area of the parabola. The same is true of a circumscribing polygon.

The proof then proceeds through contradiction: If the area of the parabola is larger than four-thirds of the triangle with the same base and height, then it is possible to increase the number of sides of the circum scribing polygon to the point where the polygon's area will be smaller than that of the parabola. If the parabola's area is smaller than that, then it is possible to increase the number of sides of the circumscribed polygon to the point where its area will be larger than that of the parabola. Both these possibilities contradict the assumption that one polygon circumscribes the parabola and that the other is circumscribed by it, and therefore the area of the parabola must be exactly four-thirds of a triangle with the same base and height. QED.

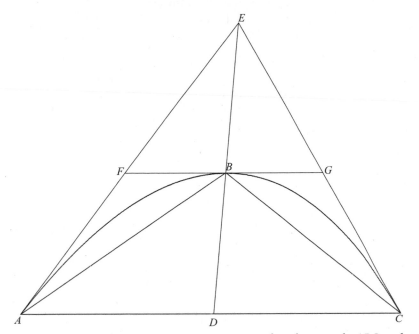

Figure 3.7. The parabola segment *ABC* circumscribes the triangle *ABC* and is circumscribed by the triangle *AEC*. As the number of sides of the polygons is increased, as in the trapezoid *AFGC*, the enclosed area more closely approximates the area enclosed by the parabola segment.

While these traditional proofs were perfectly correct, they did, Torricelli pointed out, have some drawbacks. The most obvious one is that proofs by exhaustion require one to know in advance the desired outcome—in this case, the relationship between the areas of a parabola and a triangle. Once the result was known, the method of exhaustion could show that any other relationship would lead to a contradiction, but it offered no clue as to why this relationship holds, or how to discover it. This absence led Torricelli and many of his contemporaries to believe that the ancients possessed a secret method for discovering these relationships, which they then carefully edited out of their published works. (The twentieth-century discovery of Archimedes's treatise on his nonrigorous method of discovery in the erased text of a tenth-century palimpsest suggests that they may not have been altogether wrong.)

The other chief drawback of the classical method is that it is cumbersome, requiring numerous auxiliary geometrical constructions and leading to its conclusion by a roundabout and counterintuitive route. Classical proofs, in other words, might be perfectly correct, but they were far from being useful tools for obtaining new insights.

The last ten proofs of the "De dimensione parabolae" abandoned the traditional mold of the method of exhaustion, making use of indivisibles instead. These, as Torricelli pointed out, were direct and intuitive, showing not only that the results were true, but also *why* they were true, since they were derived directly from the shape and composition of the geometrical figures in question. We have already seen how Cavalieri proved the equivalence of the two triangles making up a parallelogram by showing that they were composed of the same lines, and the equivalence of the areas enclosed by a spiral and a parabola by translating the curved indivisibles of one into the straight indivisibles of the other. Torricelli proposed the same approach for calculating the area of a parabola. The method of indivisibles, according to Torricelli, was a "new and admirable way" for demonstrating innumerable theorems by "short, direct, and positive proofs." It was "the Royal Road through the mathematical thicket," compared to which the geometry of the ancients "arouses only pity."

As Torricelli told it, the "marvelous invention" of indivisibles belonged entirely to Cavalieri, and his own contribution in the *Opera geometrica* was merely to make it more accessible. More accessible it certainly was, since Cavalieri's *Geometria indivisibilibus* was notoriously obscure, proceeding through innumerable theorems and lemmas to arrive at even the simplest results. Torricelli, in contrast, jumps directly into his mathematical problems with no rhetorical flourishes, and wastes no ink on either the verbosity or the rigor of Euclidean deduction. "We turn away from the immense ocean of Cavalieri's *Geometria*," Torricelli wrote, acknowledging the notorious difficulty of Cavalieri's text. As for him and his readers, he continued, "being less adventurous we will remain near the shore," will not bother with elaborate presentations, and will focus instead on reaching results.

Torricelli's text was so much more user-friendly than Cavalieri's that it caused considerable confusion to later generations of mathematicians. John Wallis and Isaac Barrow (1630–77) in England, and Gottfried Wilhelm Leibniz (1646–1716) in Germany, all claimed to have studied

Cavalieri and learned his method. In fact, their work clearly shows that they studied Torricelli's version of Cavalieri, believing that it was merely a clear exposition of the original. This arrangement certainly had its advantages: Torricelli, instead of defending his approach, simply refers interested readers to Cavalieri's *Geometria*, where, he assures them, they will find all the answers they seek. Later mathematicians followed his lead, and when challenged on the problematic premises of indivisibles, they, too, were happy to send their critics to seek their answers in Cavalieri's ponderous tomes.

A PASSION FOR PARADOX

In fact, there were important differences between Cavalieri's and Torricelli's approaches to infinitesimals. Most critically, in Torricelli's method all the indivisible lines taken together really did make up the surface of a figure, and all the indivisible planes actually composed the volume of a solid. Cavalieri, as will be recalled, worked hard to avoid this identification, speaking of "all the lines" as if they were different from a plane and of "all the planes" as if they were different from a solid. But Torricelli had no such qualms. In his proofs, he moves directly from "all the lines" to "the area itself" and from "all the planes" to "the volume itself," without bothering with the logical niceties that so concerned his elder. This opened Torricelli up to criticism that he was violating the ancient paradoxes on the composition of the continuum, but the truth is that Cavalieri, for all his caution, was subjected to pretty much the same critiques. At the same time, Torricelli's directness made his method far more intuitive and straightforward than Cavalieri's.

The contrast between the two is also manifest in their very different attitudes toward paradox. Cavalieri, the traditionalist, tried to avoid it at all cost, and when confronted with potential paradoxes in his method, he responded with tortured explanations of why they were not actually so. But Torricelli reveled in paradoxes. His collected works include three separate lists of paradoxes, detailing ingenious contradictions that arose if one assumed that the continuum was composed of indivisibles. This might seem surprising for a mathematician who is trying to establish the credibility of a method based precisely on this premise, but for Torricelli the paradoxes served a clear purpose. They were not merely puzzling amusements to be set aside when one engaged in serious mathematics; they were, rather, tools of investigation that revealed the true nature and structure of the continuum. The paradoxes were, in a way, Torricelli's mathematical experiments. In an experiment, one creates an unnatural situation that pushes natural phenomena to an extreme, thereby revealing truths that are hidden under normal circumstances. For Torricelli, paradoxes served much the same purpose: they pushed logic to the extreme, thereby revealing the true nature of the continuum, which cannot be accessed by normal mathematical means.

Torricelli presented dozens of paradoxes, many of them subtle and complex, but even the simplest one captures the essential problem:

> In the parallelogram ABCD in which the side AB is greater than the side BC, a diameter BD is traced with a point E along it, and EF and EG parallel to AB and BC respectively, then EF is greater than EG, and the same is true for all other similar parallels. Therefore all the lines similar to EF in the triangle ABD are greater than all the lines similar to EG in the triangle CDB, and therefore the triangle ABD is greater than the triangle CDB. Which is false, because the diameter BD divides the parallelogram down the middle.

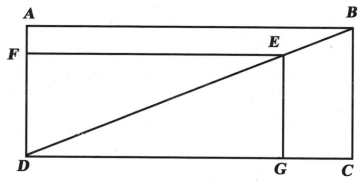

Figure 3.8. Torricelli, the paradox of the parallelogram. Based on
E. Torricelli, *Opera omnia*, vol. 1, part 2, p. 417.

The conclusion that the two halves of the rectangle differ in size is absurd, but it seems to follow easily from the concept of indivisibles.

What is to be done? Ancient mathematicians, well aware that infinitesimals could lead to such contradictions, simply banned them from mathematics. Cavalieri reintroduced indivisibles but tried to deal with such contradictions by inscribing rules into his procedures to ensure those contradictions would not arise. For example, he insisted that, in order to compare "all the lines" in one figure with "all the lines" in the other, the lines in both figures must all be parallel to a single line he called the "regula." Since the lines *EF* and *EG* in Torricelli's paradox are not parallel, then Cavalieri could claim that they should not be compared at all, and the paradox can be averted. In practice, however, Cavalieri's artificial limitations were ignored both by his followers, who saw them as inconvenient hindrances, and by his critics, who did not believe that they resolved the fundamental problem.

Torricelli took a different approach. Instead of trying to evade the paradox, he made a sustained effort to understand it and what it meant for the structure of the continuum. His conclusion was startling: The reason all the short lines parallel to *EG* produce an area equal to the same number of longer lines parallel to *EF* is that the short lines are "wider" than the long lines. More broadly, according to Torricelli, "that indivisibles are all equal to each other, that is that points are equal to points, lines are equal in width to lines, and surfaces are equal in thickness to surfaces, is an opinion that seems to me not only difficult to prove, but in fact false." This is a stunning idea. If some indivisible lines are "wider" than others, doesn't that mean that they can in fact be divided, to reach the width of the "thin" lines? And if indivisible lines have a positive width, doesn't it follow that an infinite number of them would add up to an infinite magnitude—not to the finite area of the triangles *ADB* and *CDB*? And the very same applies to points with a positive size and surfaces with a "thickness." The assumption seems absurd, but Torricelli insisted that his paradoxes indicated that there was no other explanation. And not only that: he founded his entire mathematics approach on precisely this idea.

In order to transform this basic insight into a mathematical system, it was not enough to say in principle that indivisibles differed in size from one another; it was necessary to determine by precisely how much they differed from one another. For this, Torricelli turned once again to the paradox of the parallelogram. In the diagram, the same number of long lines *EF* and short lines *EG* produce exactly the same total area. For this to be true, the short lines *EG* need to be "wider" by exactly the same proportion as the lines *EF* are "longer." That in turn is the ratio of *BC* to *BA*, which is, in other words, the slope of the diagonal *BD*. At a stroke, Torricelli transformed a rather dubious speculation about the composition of the continuum into a quantifiable and usable mathematical magnitude.

Torricelli then showed exactly how to make mathematical use of indivisibles with "width" by calculating the slope of the tangent of a class of curves that we would characterize as $y^m = kx^n$, and that he called an "infinite parabola." In this he went well beyond Cavalieri, who calculated areas and volumes enclosed in geometrical curves, but never their tangents. Indeed, Cavalieri's insistence on comparing only conglomerations of "all the lines" or "all the planes" left no room for the delicate calculation of tangents, which are slopes calculated at single indivisible points. But Torricelli's more flexible method, which distinguished between the magnitudes of different indivisibles, made this possible. He first pointed to the figures *ABEF* and *CBEG* in the paradox of the parallelograms. The two figures, known as "semi-gnomons," are equal in area because they complete the equal triangles *DFE* and *EGD* to the equal triangles *ADB* and *CDB*. This will always be true, furthermore, no matter where the point *E* is positioned on the diagonal *DB*, even as it is moved to the point *B* itself. Accordingly, the line *BC* is equal in area, or "quantity," to the line *AB*, even though the line *AB* is longer. This is the case because, just like the semi-gnomon *CBEG*, the indivisible line *BC* is "wider" than *AB* by precisely the same ratio that *AB* is longer.

Now, as long as we are dealing with straight lines, such as the diagonal *BD*, the semi-gnomons are always equal, and the "width" of the indivisibles is given by the simple ratio of the slope. But what happens if, instead of a straight line, we are given a generalized parabola, which in modern terms would be given as $y^m = kx^n$? In this "infinite parabola," the semi-gnomons are no longer equal, but they do hold a fixed relationship. As Torricelli proved using the classic method of exhaustion, if the segment on the curve is very small, the ratio of the two semi-gnomons is as $\frac{m}{n}$. And if the width of the semi-gnomons is only a single indivisible, then the "size" of the indivisible lines that meet at the curve is as $\frac{m}{n}$.

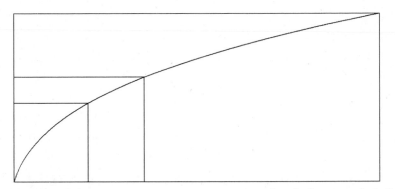

Figure 3.9. Semi-gnomons meeting at a segment of an "infinite parabola."
If the segment is very small, or indivisible, then the ratio of the areas of
the semi-gnomons is as $\frac{m}{n}$.

This result enabled Torricelli to calculate the slope of the tangent at every point on the "infinite parabola," shown as the curve AB in figure 3.10. Torricelli's key insight is that at the point B, where the two indivisible lines BD and BG meet the curve, they also meet the straight line that is the curve's tangent at that point. And whereas the "area" of the two indivisibles is as $\frac{m}{n}$ relative to the curve, it is equal relative to the straight tangent—if the tangent is extended to become the diagonal of a rectangle. Accordingly, in Figure 3.10, the ratio of the "areas" of BD and BG is $\frac{m}{n}$, but the ratio of the "areas" of BD and BF is 1. This means that the ratio of the "areas" of BF and BG is $\frac{m}{n}$. Now, BF and BG have the same "width" in Torricelli's scheme, because they both meet the curve BF (or its tangent) at point B at precisely the same angle. The difference between the two segments is only in their lengths, and it follows that the length BF is to the length BG as $\frac{m}{n}$. Now, BF is equal to ED, and BG is equal to AD, and therefore the ratio of the abscissa ED of the tangent to the abscissa AD of the curve is $\frac{m}{n}$, or, more simply, $ED = \frac{m.AD}{n}$. Therefore the slope $\frac{BD}{ED}$ of the tangent at point B is $\frac{n.BD}{m.AD}$. In this way, the slope of an "infinite parabola" can be known at any given point on the "infinite parabola," based on its abscissa and ordinate.

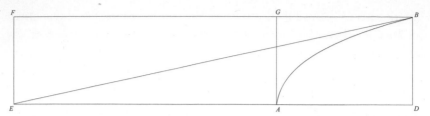

Figure 3.10. Torricelli's calculation of the slope of an "infinite parabola."

The significance of Torricelli's procedure here extends beyond the ingenuity of the proof itself (which is considerable), and to the challenge it posed to the mathematical tradition. Since ancient times, mathematicians had shied away from paradoxes, treating them as insurmountable obstacles, and a sign that their calculations had reached a dead end. But Torricelli parted ways with this venerable tradition: instead of avoiding paradoxes, he sought them out and harnessed them to his cause. Galileo had speculated about the infinitesimal structure of the continuum, but qualified his remarks by admitting that the continuum was a great "mystery." Cavalieri did his best to avoid paradoxes and to conform to traditional canons, even at the cost of making his method unwieldy. But Torricelli unapologetically used paradoxes to devise a precise and powerful mathematical tool. Instead of banishing the paradox of the continuum from the realm of mathematics, Torricelli placed it at the discipline's heart.

Despite its clear logical perils, Torricelli's method made a profound impression on contemporary mathematicians. Although constantly skirting the edges of error, it was also flexible and remarkably effective. In the hands of a skilled and imaginative mathematician, it was a powerful tool that could lead to new and even startling results. In the 1640s it spread quickly to France, where it was developed by the likes of Gilles Personne de Roberval and Pierre de Fermat (1601–65), who corresponded directly with Torricelli. The Minim father Marin Mersenne, who was the central node of the European "Republic of Letters," also corresponded with Torricelli, and then spread the Italian's method on to England, where Wallis and Barrow mistakenly attributed it to Cavalieri. Quickly disseminating across the Continent, Torricelli's radical practice encapsulated the power and the promise, as well as the dangers, of the new infinitesimal mathematics.

Torricelli did not enjoy his newfound prominence for long. On October 5, 1647, he fell ill, and less than three weeks later, on October 25, he was dead at the age of thirty-nine. In an hour of lucidity shortly before his death, Torricelli instructed his executors to deliver his manuscripts to Cavalieri in Bologna, so that he could publish what he saw fit. But it was too late: on November 30, just over a month after Torricelli breathed his last, Cavalieri, too, was dead from the gout that had afflicted him for many years. Within a few short years Italian mathematics was deprived of its guiding light Galileo and of his two chief mathematical disciples. In the span of a few decades these three had transformed the face of mathematics, opening up new avenues of progress, and possibilities that were eagerly seized upon by mathematicians across Europe. A generation later their "method of indivisibles" would be transformed into Newton's "method of fluxions" and Leibniz's differential and integral calculus.

In their own land, however, Galileo, Cavalieri, and Torricelli would have no successors. For just as Italian mathematics was being deprived of the leadership of Galileo and his disciples, the tide in Italy was turning decisively against their brand of mathematics. The Society of Jesus, which had long viewed the method of indivisibles with suspicion, had swung into action. In a fierce decades-long campaign, the Jesuits worked relentlessly to discredit the doctrine of the infinitely small and deprive its adherents of standing and voice in the mathematical community. Their efforts were not in vain: as 1647 was drawing to a close, the brilliant tradition of Italian mathematics was coming to an end as well. It would be centuries before the land of Galileo, Cavalieri, and Torricelli was once again home to creative mathematicians of the highest rank.

4

"Destroy or Be Destroyed": The War on the Infinitely Small

THE DANGERS OF THE INFINITELY SMALL

The Jesuit mathematician André Tacquet (1612–60) was, by the standards of his time, a man of the world. Although he may never have left his native Flanders, his network of correspondents spanned Europe's religious divide, reaching to Italy and France, but also to Protestant Holland and England. Only months before his death he entertained the Dutch polymath Christiaan Huygens, who had traveled to Antwerp with the express purpose of meeting Tacquet, by then regarded as one of the brightest mathematical stars ever to come out of the Society of Jesus. The two spent only a few days together, but got along so well that the Jesuit was convinced that he had managed to lure Huygens to the Catholic faith. (He hadn't.) But ultimately it was not Tacquet's personal charm, but rather his mathematical excellence that transcended seventeenth-century prejudices. In England, Henry Oldenburg, secretary of the Royal Society of London and no friend of the Jesuits, spent so much time describing Tacquet's *Opera mathematica* at the Society's meeting in January 1669 that he felt compelled to apologize to the fellows for abusing their patience. But it was, he insisted, "one of the best books ever written on mathematics."

Tacquet's claim to mathematical fame rested chiefly on his 1651 book *Cylindricorum et annularium libri IV* ("Four Books on Cylinders and Rings"), in which he showed a complete mastery of the full mathematical arsenal available in his day. He calculated the areas and volumes of geometrical figures using both classical approaches and the new methods developed by his contemporaries and immediate predecessors. But when it came to indivisibles, the usually mild-mannered Jesuit turned blunt:

> I cannot consider the method of proof by indivisibles as either legitimate or geometrical . . . many geometers agree that a line is generated by the movement of a point, a surface by a moving line, a solid by a surface. But it is one thing to say that a quantity is generated from the movement of an indivisible, a very different thing to say that it is *composed* of indivisibles. The truth of the first is altogether established; the other makes war upon geometry to such an extent, that if it is not to destroy it, it must itself be destroyed.

Destroy or be destroyed—such were the stakes when it came to infinitesimals, according to Tacquet. Strong words indeed, but to the Fleming's contemporaries, they were not particularly surprising. Tacquet was, after all, a Jesuit, and the Jesuits were then engaged in a sustained and uncompromising campaign to accomplish precisely what Tacquet was advocating: to eliminate the doctrine that the continuum is composed of indivisibles from the face of the earth. Should indivisibles prevail, they feared, the casualty would be not just mathematics, but the ideal that animated the entire Jesuit enterprise.

When Jesuits spoke of mathematics, they meant Euclidean geometry. For, as Father Clavius had taught, Euclidean geometry was the embodiment of order. Its demonstrations begin with universal self-evident assumptions, and then proceed step by logical step to describe fixed and necessary relations between geometrical objects: the sum of the angles in a triangle is always equal to two right angles; the sum of the squares of the two shorter sides of a right-angled triangle is equal to the square of the long side; and so on. These relations are absolute, and cannot be denied by any rational being.

And so, beginning with Clavius and for the next two hundred years, geometry formed the core of Jesuit mathematical practice. Even in the eighteenth century, when the direction of higher mathematics turned decisively away from geometry and toward the newer fields of algebra and analysis, Jesuit mathematicians held firm to their geometrical practice. It was the unmistakable hallmark of the Jesuit mathematical school. If only theology and other fields of knowledge could replicate the certainty of Euclidean geometry, they believed, then surely all strife would be at an end. The Reformation and all the chaos and subversion that flowed from it would never have taken root in such a world.

This vision of eternal order was, to the Jesuits, the only reason mathematics should be studied at all. Indeed, as Clavius never tired of arguing to his skeptical colleagues, mathematics embodied the Society's highest ideals, and thanks to his efforts the doors were opened at Jesuit institutions for the study and cultivation of the field. By the late sixteenth century, mathematics had become one of the most prestigious fields of study at the Collegio Romano and other Jesuit schools.

Just as Euclidean geometry was, for the Jesuits, the highest and best of what mathematics could be, so the new "method of indivisibles" advocated by Galileo and his circle was its exact opposite. Where geometry began with unassailable universal principles, the new approach began with an unreliable intuition of base matter. Where geometry proceeded step by irrevocable step from general principles to their particular manifestations in the world, the new methods of the infinitely small went the opposite way: they began with an intuition of what the physical world was like and proceeded to generalize from there, reaching for general mathematical principles. In other words, if geometry was top-down mathematics, the method of indivisibles was bottom-up mathematics. Most damaging of all, whereas Euclidean geometry was rigorous, pure, and unassailably true, the new methods were riddled with paradoxes and contradictions, and as likely to lead one to error as to truth.

If infinitesimals were to prevail, it seemed to the Jesuits, the eternal and unchallengeable edifice of Euclidean geometry would be replaced by a veritable tower of Babel, a place of strife and discord built on teetering foundations, likely to topple at any moment. If Euclidean geometry was, for Clavius, the foundation of universal hierarchy and order, then the new mathematics was the exact opposite, undermining the very possibility of universal order, leading to subversion and strife. Tacquet was not exaggerating when he wrote that in the struggle between geometry and indivisibles, one must destroy the other or "must itself be destroyed." And so the Jesuits proceeded to do just that.

THE CENSORS, PART I

The issue of the structure of the continuum could hardly have been further removed from the minds of the early Jesuit fathers as they faced off against Martin Luther and his followers in a battle for the soul of Europe. The first Jesuit to take any notice of the issue was none other than Clavius's old nemesis at the Collegio Romano, Benito Pereira. In 1576, at the height of his struggle with Clavius over the proper place of mathematics in the Jesuit curriculum, Pereira published a book on natural philosophy intended to establish the proper principles that should be adopted by the Jesuits. Following the guidelines established by the Society's founders, Pereira adhered closely to the teachings of Aristotle, and so he also addressed that ancient philosopher's teachings on the subject of the continuum. In the best tradition of medieval scholasticism, Pereira first posed the thesis that a line is composed of separate points, and presented all the arguments offered in support of the thesis by ancient and medieval masters. He then demolished the arguments one by one, until he was left to conclude, along with Aristotle, that the continuum is infinitely divisible, and not composed of indivisibles. Pereira, it is clear, was not concerned about mathematical innovations or their subversive implications: writing decades before Galileo and his disciples developed their radical mathematical techniques, he had no reason to be. And since he saw no value for the Jesuits in the study of any kind of mathematics, he was unlikely to concern himself with determining the right "kind" of mathematics that should be taught. For him the question of the continuum was merely one more topic to be addressed in a discussion of Aristotle's natural philosophy.

It was a full two decades before another Jesuit took up the question of the continuum, and this time it was a much more authoritative one: Father Francisco Suárez, the leading theologian of the Society of Jesus. In 1597, Suárez devoted thirteen folios to the question of the composition of the continuum in his *Disputation on Metaphysics*, but like Pereira, he addressed the matter as part of a broader discussion of Aristotelian physics. Unlike Pereira, however, the great theologian did not peremptorily reject the notion that the continuum is composed of indivisibles; admitting that the question is difficult, he gives up any hope of certainty, and seeks only an answer that "appears to be true." He cites the doctrine that the continuum is composed of indivisibles, and then the complete denial of indivisibles, arguing that both are "extreme" positions. He then proposes some intermediate positions that he thinks more likely, while conceding that a definite solution is beyond reach. For Suárez as for Pereira, the entire question was technical, or what we would call "academic." Neither one thought that there was much at stake here, except the correct interpretation of Aristotelian physics.

But as the troubled century of Charles V, Luther, and Ignatius was drawing to a close, an unmistakable sense of urgency entered into Jesuit discussions of infinitesimals. At the time, the general superior, Father Claudio Acquaviva, was increasingly concerned with the growing diversity of opinions within the Society. This was undoubtedly the price of success, as the rapid expansion of the Society in those years, in the form of hundreds of colleges and missions across the known world, brought many new peoples into its orbit. But for General Acquaviva this was no excuse for the soldiers of Christ to deviate from the correct teachings of the Church. As far as the Jesuit hierarchy was concerned, the increase in the Society's numbers and influence was all the more reason it should speak with a single and clear voice. "Unless minds are contained within certain limits," warned Father Leone Santi, prefect of studies at the Collegio Romano some years later, "their excursions into exotic and new doctrines will be infinite," leading to "great confusion and perturbation to the Church." To prevent this, the general superior in 1601 instituted a college of five Revisors General at the Collegio Romano, with the power to censor anything that was taught in the Society's schools anywhere in the world or published under the Society's aegis. With the Revisors' oversight, Acquaviva hoped, only correct doctrine would be taught in Jesuit schools, and books published by Jesuit fathers would speak with a single authoritative voice, approved by the authors' superiors. It did not take long for the Revisors to begin issuing prohibitions on the teaching and promotion of infinitesimals.

The first decree by the Revisors General on the composition of the continuum dates from 1606, when the office was only five years old. Responding to a proposition sent in from the Society's schools in Belgium that "the continuum is composed of a finite number of indivisibles," the Revisors, quickly and without comment, ruled that the proposition was an "error in philosophy." Only two years later another missive from

Belgium brought the very same doctrine before the Revisors. This time they were somewhat more expansive, though just as firm: "everyone agrees that this must not be taught, since it is improbable and also certainly false and erroneous in philosophy, and against Aristotle." Only a decade before, Suárez had merely expressed some concern about whether the notion that the continuum was composed of indivisibles was philosophically viable, and offered some alternatives. The Revisors, in contrast, banned it outright as "false and erroneous."

What had changed? The Revisors themselves offer no clue, and the summaries they left provide no details on the sources of propositions brought before them, except for their country of origin. But we do know that those early years of the seventeenth century saw a significant uptick in interest in the infinitely small among mathematicians. In 1604, Luca Valerio of the Sapienza University in Rome published a book on calculating the centers of gravity of geometrical figures in which he employed rudimentary infinitesimal methods. Valerio was well known to the Jesuits, having studied under Clavius for many years, and even receiving doctoral degrees in philosophy and theology from the Collegio Romano. His work could not have gone unnoticed by the Jesuit fathers, who likely felt they needed to better define their position on this new approach. We also know that in 1604, Galileo, then at the University of Padua, was experimenting with indivisibles in formulating his law of falling bodies. Galileo thought very highly of Valerio: years later he nominated him for membership in the Lincean Academy, and in his *Discourses* of 1638 he refers to Valerio as "the Archimedes of our age." Whether the two drew upon each other or developed their ideas independently, their work marked a significant change in the status of infinitesimals: instead of an ancient doctrine definitively discussed by Aristotle and his later commentators, infinitesimals now seemed to be entering the arena of contemporary mathematics.

For the Jesuits, this was a critical change. Clavius had only recently won his battle to establish mathematics as a core discipline in the Jesuit curriculum, and the order's mathematicians were beginning to be recognized as leaders in the field. When, in the early seventeenth century, infinitesimals began seeping into mathematical practice, the Jesuits felt compelled to take a stand on the new methods. Are they compatible with the Euclidean approach so central to the Society's mathematical practice? The Revisors' answer was a resounding no. Yet despite the stern pronouncements, the problem never seemed to go away. Mathematically trained Jesuits across Europe were closely following developments at the forefront of mathematical research, and were well aware of the growing interest in infinitesimals. Conscious of the sensitivity of the subject, they kept appealing to the Revisors with different versions of the doctrine, each one deviating slightly from those that had already been banned. Consequently, when the Revisors turned their attention once more to the infinitely small, the catalyst once again was developments in the field of mathematics.

Johannes Kepler (1571–1630) is remembered today as the man who first plotted the correct elliptical paths of the planets through the heavens. Nor did Kepler go unappreciated in his own day. In the early seventeenth century he was the only scientist in the world whose fame rivaled Galileo's, and though a Protestant, he held the most coveted mathematical position in the world: court astronomer to the Holy Roman Emperor in Prague. In 1609 Kepler published his masterpiece *Astronomia nova* (*The New Astronomy*), in which he demonstrates that the planets move in ellipses and not perfect circles, and codifies his observations in two laws of planetary motion. (Kepler's third law was published later, in his *Harmonices mundi* of 1619.) To calculate the precise motion of the planets at varying speeds along their orbit, Kepler made rough use of infinitesimals, assuming that the arc of their elliptical path was composed of an infinite number of points. Six years later Kepler further developed his mathematical theory in a work dedicated to calculating the exact volume of wine casks, where he calculated a whole range of areas and volumes of geometrical figures using infinitesimal methods. To calculate the area of a circle, for example, he assumed that it was a polygon with an infinite number of sides; a sphere was composed of an infinite number of cones, each with its tip at the center and its base on the surface of the sphere, and so on. Titled *Nova stereometria doliorum vinariorum* ("A New Stereometry of Wine Caskets"), it was a mathematical tour-de-force that hinted at the power of the approach that Cavalieri would later systematize and name. Once again the Jesuits felt compelled to respond, and once again the job fell to the Revisors General in Rome. In 1613 they denounced the proposition that the continuum was composed of either physical "minims" or mathematical indivisibles. In 1615 they reiterated their condemnation, rejecting, first, the opinion that "the continuum is composed of indivisibles" and, several months later, the opinion that "the continuum is composed of a finite number of indivisibles." This doctrine, they opined, "is also not permitted in our schools . . . if the indivisibles are infinite in number."

Once the Revisors had issued their decision, a well-oiled machinery of enforcement sprang into action. The numerous Jesuit provinces across the globe were informed of the censors' verdict, and they then passed it on to lower and then lower jurisdictions. At the end of this chain of transmission were the individual colleges and their teachers, who were instructed on the new rules on what was permissible and what was not. Once a decision by the Revisors in Rome descended the Jesuit hierarchy and reached an individual professor, he was now responsible for carrying it out to the fullness of his ability and of his own free will, regardless of his previous views on the subject. It was a system based on hierarchy, training, and trust—or, as an unfriendly observer might suggest, on indoctrination. Either way, there was no doubt that it was remarkably effective: the Revisors' pronouncements became law in the many hundreds of Jesuit colleges worldwide.

THE FALL OF LUCA VALERIO

The Revisors' decree of 1615 against infinite indivisibles may have been directed against the admirers of Kepler. But whatever the intent, it was the Jesuits' former associate Luca Valerio who fell victim to the Society's new and harsher stance. Three years had passed since Galileo proposed Valerio for membership in the prestigious Lincean Academy, which served as the institutional center of the Galileans in Rome. The academy was an exclusive club made up of a select group of leading scientists and their aristocratic patrons, but Valerio seemed like a perfect fit: not only was he a mathematician renowned for bold ideas and a professor at the ancient Sapienza University, but he was also an aristocrat and a personal friend of the late pope Clement VIII (1592–1605), who had been his student. He brought with him sparkling social prestige, as well as personal creativity and institutional respectability, and the Linceans promptly elected him on June 7, 1612. From the moment of his election, Valerio became a leader among the Linceans, given overall editorial responsibility for all the academy's publications.

Valerio, who had studied for years under Clavius, remained on good terms with his former mentors and colleagues at the Collegio Romano, and that, too, made him valuable to the Linceans. At a time of increasing tensions between the Galileans and the Jesuits of the Collegio Romano, Valerio served as a means of communication and possible compromise between the two camps. Indeed, there was nothing Valerio wanted more than to heal the rift that had opened between his two groups of friends. It was not to be. Showing no concern for Jesuit sensitivities, Galileo published his *Discourse on Floating Bodies*—which attacked the principles of Aristotelian physics—debated the nature of sunspots, and circulated his views on the proper interpretation of Scripture. For the Jesuits of the Collegio Romano, this intrusion into theology was the last straw. They determined to strike back against the man they had once honored with a full day of ceremonies but whom they now viewed as a bitter enemy.

The Jesuits had learned from their past mistakes. Time and again they had been outmatched by Galileo's brilliant polemics, coming off as rigid and didactic pedants standing in the way of scientific progress. So, instead of engaging in public debate, they turned to the arena in which their power was unchallengeable: the hierarchy and authority of the Church. In 1615 cardinal Bellarmine issued his opinion against Copernicanism, which soon became official Church doctrine. He followed this up with a personal warning to Galileo to desist forever from holding or advocating the forbidden doctrine. It was an impressive demonstration of the Jesuits' ability to harness the Church apparatus to their cause, and a stinging defeat to the Galileans. As regards infinitesimals, nothing so public as the decree against Copernicanism took place. But it is probably not a coincidence that the Revisors' ruling against indivisibles in April 1615 coincided precisely with Bellarmine's issuing his opinion against Galileo.

Valerio felt besieged. The two great intellectual schools that he had hoped to reconcile were now openly at war. The middle ground on which he stood was quickly melting away, and he was being pulled in opposite directions. The Revisors' decree of April 1615 on the composition of the continuum was a reminder that, as a mathematician identified with infinitesimal methods, he could not long remain above the fray. When the

Revisors repeated their decree in November, this time adding that it applied even "if the indivisibles are infinite in number," he may well have concluded that he himself was their target. We do not know what he was told in private by either the Jesuits or the Linceans, but the pressure must have been unbearable. Finally, in early 1616, with the tide turning decisively against the Galileans, Valerio made his decision: he tendered his resignation to the Lincean Academy, siding openly with the Jesuits.

The Linceans were stunned. Membership in the academy was a great honor, and never before had anyone turned his back on it. That such a thing could happen at all was a measure of just how precarious the Galileans' position had become in the face of the Jesuit onslaught. Undeterred, the Linceans responded decisively: they promptly refused Valerio's resignation, on the grounds that it conflicted with the oath taken by every member of the academy. Valerio therefore remained technically a Lincean, but only in name: in a meeting on March 24, 1616, his fellows censored him for betraying his oath of loyalty and offending both Galileo and "Lyncealitas," the Lincean principle of mutual solidarity. They then barred Valerio from any future meetings of the academy and, for good measure, deprived him of his voting rights.

Valerio had misread the signs. The Galileans may have been put on the defensive, but they were still powerful enough to strike back at their former colleague. His life and career, so brilliantly successful for so long, ended up a Greek tragedy. Recognized early for his mathematical prowess, he had scaled the heights of scholarly fame in Italy, admired and honored by both conservatives and innovators. But when he could no longer bridge the growing divide between the two, he made a choice, and it turned out to be the wrong one. Isolated, humiliated, and a pariah to his former friends, Valerio retired, and died less than two years after his expulsion by the Linceans, an early victim of the Jesuit war against the infinitely small.

GREGORY ST. VINCENT, SJ

Valerio had studied and trained at the Collegio Romano for many years, but he was not a Jesuit himself. Sometimes, however, Jesuit officials had to deal not with outsiders but with their own members, Jesuit intellectuals

who pushed back against the strictures placed upon them by their superiors, trying their best to pursue their work freely while still adhering to
the letter, if not the spirit, of the law. In such cases the Jesuits usually
took a milder approach, preferring to remind a wayward member of the
common Jesuit bond and the ideal of voluntary obedience. By relying on
their hierarchical order and the deeply ingrained value of obedience, the
Jesuits managed to exert much greater control over their members than
they would likely have achieved through disciplinary action, force, or
intimidation.

Even so, challenging the order's decrees had its price, as the mathematician Gregory St. Vincent of Bruges found to his dismay. A Fleming,
like his younger contemporary Tacquet, St. Vincent (1584–1667) was one
of the most creative mathematical minds ever to come out of the Society
of Jesus. In 1625, while teaching at the Jesuit College in Louvain, St. Vincent developed a method for calculating areas and volumes of geometrical
figures, which he called "ductus plani in planum." His greatest triumph,
he believed, was solving an ancient problem that had stumped the
greatest geometers of every age: constructing a square equal in area to a
given circle, or more simply, "squaring the circle." St. Vincent decided to
publish his results, and good Jesuit that he was, he sent his manuscript to Rome to obtain permission. Since St. Vincent was a distinguished
mathematician, and his text technical and challenging, his request made
it up through the ranks all the way to Mutio Vitelleschi, the general superior of the Society of Jesus.

Vitelleschi hesitated: most mathematicians of the era believed (rightly,
as it turned out) that squaring the circle was impossible, or at least impossible by classical Euclidean methods. Those who claimed to have
accomplished the feat were usually dismissed as quacks, and there was
a significant danger that a Jesuit scholar who claimed he had succeeded
in squaring the circle would tarnish the Society's reputation. More troubling was the fact that St. Vincent's "ductus plani in planum" method
looked suspiciously like it was based on the forbidden doctrine of infinitesimals. Not wishing to decide a technical matter himself, Vitelleschi
passed the matter to Father Grienberger, Clavius's student and successor at the Collegio Romano, and the highest mathematical authority in
the order. Grienberger read the treatise closely, but he, too, was unconvinced, and ruled against publication. Undeterred, St. Vincent requested

and received permission to travel to Rome, where for two years he tried to convince Grienberger that his method was valid and did not violate the strictures against infinitesimals. He failed. In a 1627 letter, Grienberger informed Vitelleschi that while he did not doubt the correctness of St. Vincent's results, he nevertheless had serious concerns regarding his method. The Fleming returned to Louvain empty-handed, and published nothing for the next twenty years. Only in 1647, taking advantage of the recent death of General Vitelleschi, did St. Vincent finally bring his work to print. This time, circumventing the authorities in Rome, he settled for a curt permission by the Jesuit provincial of Flanders, allowing the work to be printed.

St. Vincent's experience typifies the Jesuit attitude toward indivisibles in the years that followed the prohibitions of 1615 and Valerio's fall from grace. Indivisibles were prohibited, to be sure, but policing their use was not a high priority for the Society. When confronted with new mathematical methods that made use of the infinitely small, the Jesuits took action, reminding their members that these approaches were not permitted. Beyond that, however, they did little to curb the spread of the new approach. The fact that a distinguished Jesuit such as St. Vincent had developed a method inspired by indivisibles, and that he sent it to his Roman superiors to be approved, attests to the fact that he expected to be allotted some wiggle room. His application to publish his work was denied, but beyond that, St. Vincent was not punished for his transgression. He continued to hold distinguished positions in Jesuit colleges for the rest of his long life and, taking advantage of an opportune moment, even managed in the end to publish his work. In later years, when the Jesuits pursued the infinitely small with an air of grim finality, they would not be so forgiving.

THE ECLIPSE OF THE JESUITS

St. Vincent, as it turns out, was fortunate in his timing, which fell in the midst of a lull in the Jesuit campaign to curb infinitesimals. After the prohibition of 1615, the Revisors did not return to the issue for another seventeen years. The reason had much to do with the changing fortunes of the Society itself. Following their victory over the Galileans in 1616, the Jesuits reigned supreme in Rome. Pope Paul V had sided with them publicly in the fight with Galileo, humbling their critics and establishing them as the arbiters of truth. The advocates of Copernicanism had been effectively silenced, as had, it seemed, any talk of the infinitely small. Valerio had been humiliated, and Galileo was in no position to challenge the authority of the Collegio Romano. In 1619, Pope Paul V showed his favor to the Jesuits by beatifying Ignatius's acolyte, the intrepid missionary Francis Xavier, just as he had done for Ignatius himself ten years before. In 1622, Paul's successor, Pope Gregory XV, completed the process, making St. Ignatius and St. Francis Xavier the first Jesuit saints of the Catholic Church. In celebration, the Jesuits began plans for construction of a magnificent new Church of St. Ignatius on the grounds of the Collegio Romano.

But in the autocracy that was the papal Curia, the one source of authority and power that mattered was the personal favor of the Pope. This was rarely a problem for the Jesuits, who championed papal supremacy in the Church, and whose elite swore a personal oath of obedience to the Pope. Not surprisingly, most popes thought it in their best interest to bestow their favor on the Society. Nevertheless, there were exceptions: Pope Paul IV (1555–59), for example, a founder of the rival order of the Theatines, was hostile to the Jesuits, and during his reign the Society suffered. Now, seven decades later, this history seemed to be repeating itself. In July of 1623, only two years after his election, Pope Gregory XV died, throwing all political calculations in Rome into chaos. It took a month of intense maneuverings for the College of Cardinals to decide on a successor, but when the dust finally settled, it was clear to all that a new day had dawned in Rome: the man elected was the Florentine cardinal Maffeo Barberini, who chose the name of Urban VIII. For the Jesuits, the choice could hardly have been worse.

There were many reasons the Jesuits would view the election of Barberini with apprehension. For one thing, he was from Florence, a city that prided itself on its tradition of independence and where, consequently, the Jesuits held relatively little sway. It was, after all, the protection of Florence's ruler, Grand Duke Cosimo II, that had saved Galileo from far more severe consequences in his tangle with the Jesuits in 1616. Barberini had also served for many years as papal nuncio to France, and was known to be close to the French court. Indeed, it was French influence, in the person of Cardinal Maurizio of Savoy, that had secured Barberini's election as Pope. The Jesuits, in contrast, tangled regularly with the French monarchy and its advocates at the Sorbonne over the issue of papal supremacy, and on more than one occasion were banned from teaching in France for refusing to take an oath of obedience to the king. At the papal Curia, the Jesuits were staunch supporters of France's rivals, the Habsburgs—Holy Roman emperors and kings of Spain— whom they viewed as the best hope for restoring the unity of Christendom. But perhaps most troubling to the Jesuits was the fact that Barberini was a personal friend of Galileo, and had openly professed his admiration for his fellow Florentine, his discoveries, and his opinions. In the Roman culture wars of the previous decade, Barberini had come down firmly on the side of Galileo and the Lincean Academy, the enemies of the Jesuit fathers at the Collegio Romano.

No sooner had Urban VIII settled in his office than he began behaving in ways that confirmed the Jesuits' worst fears. He appointed Monsignor Giovanni Ciampoli as his personal secretary, and the young duke Virginio Cesarini as master of the papal secret chamber. Both men were Linceans who had plotted with Galileo on how to "bring down the pride of the Jesuits." Following Cardinal Bellarmine's death in 1621 the Jesuits were left without a representative in the College of Cardinals, and the new Pope seemed happy to leave things that way. When in 1627 the Jesuits petitioned him to have Bellarmine elevated to sainthood, Urban was in no hurry to respond. Instead, he set up a new barrier in the process, ruling that fifty years must pass between a candidate's death and his canonization. But most troubling to the Jesuits was that Urban VIII's admiration for Galileo and his eagerness to speak about it had not diminished with his accession. In 1623, when Galileo published *The Assayer*, his latest and most powerful salvo in his war of words with the Collegio Romano, the Pope welcomed it warmly. He publicly accepted a specially bound copy of the book, personally dedicated by Galileo, from Prince Federico Cesi, founder of the Linceans, and asked Ciampoli to read it to him at table. As Cesarini, who was present at these occasions, assured Galileo, the new Pope was amused and full of admiration. Taking advantage of the favorable constellation, the Linceans quickly inducted the Pope's nephew Francesco Barberini into their academy, a move whose wisdom was soon confirmed when the Pope bestowed the cardinal's purple on young Francesco. Now the cardinal nephew, perhaps the second most powerful man in Rome, was himself a Lincean, while the Jesuits went for decades without a cardinal of their own.

Nor did the new Pope's foreign policy endear him to the Jesuits. Indeed, Urban VIII was something of a throwback to the Renaissance popes, concerned more with securing his independence as an Italian prince than with enhancing his standing as the spiritual leader of all Christians. In the midst of the Thirty Years' War (1618–48), the bloodiest of all religious struggles, the Pope was naturally expected to side firmly with the Habsburgs, whose imperial domains were being ravaged and who bore the brunt of the struggle against Protestantism. Instead, Urban tried to free himself from the suffocating Habsburg embrace by allying himself with France and its enigmatic chief minister, Cardinal Richelieu. Rather than support the war on the heretics, Urban built up his own military power and bullied neighboring Italian potentates—all good Catholics. He added the Duchy of Urbino to his domains, becoming the last Pope to expand the Papal States, and launched the "wars of Castro" against the Farnese dukes of Parma. In 1627, when there was no male heir to the ancient Gonzaga line of Mantua, he actually supported the succession of a Protestant, Charles Gonzaga, Duke of Nevers, over the Habsburg claimant.

Even in their season of discontent, the Jesuits, ever pious, ever active, tried to direct Church policy away from what they considered a disastrous course. Their plan hinged on a perennial bone of contention between Rome and Paris: the question of papal supremacy. In 1625, Antonio Santarelli, professor at the Collegio Romano, authored a spirited defense of papal power in a book with the imposing title *Treatise on Heresy, Schism, Apostasy, the Abuse of the Sacrament of Penance, and the Power of the Roman Pontiff to Punish These Crimes*. His main thesis was that popes reigned supreme over secular monarchs, and even had the power to remove them if they acted in ways harmful to the faith. This doctrine was hardly new, and seemed quite self-evident to the Jesuits, who believed in a strict hierarchy of Church, state, and society, with the Pope at its apex. Back in 1610, Cardinal Bellarmine himself had published his *Treatise on the Power of the Supreme Pontiff in Temporal Affairs*, making much the same claims on behalf of papal power. But as the cardinal found out, what was a simple truism in Rome was sedition in Paris, where the Bourbons were busy building an absolute monarchy in which royal authority would reign uncontested. Bellarmine's book was publicly condemned by the Parlement of Paris, the Jesuits were banned from teaching in France for several years, and a brief diplomatic crisis flared between the Bourbon court and the Vatican. By publishing Santarelli's book in 1625 in Paris, the Jesuits were likely hoping to instigate a similar crisis, which would force the Pope, willingly or not, back into the arms of the Habsburgs.

The plot misfired badly. The French were indeed incensed with Santarelli's treatise, but they focused their fury strictly on the Jesuits rather than on the Pope. The book was condemned to the flames by the Parlement of Paris, and denounced by the faculty of the Sorbonne and other French universities. On March 16, 1626, the leaders of the French Jesuits were called before the Parlement and asked to sign a public disavowal of Santarelli's "evil doctrine." Faced with the destruction of their entire French mission if they refused, the Jesuits humbly signed. If this was not embarrassment enough, a greater one awaited them in Rome: on May 16 the Pope summoned Mutio Vitelleschi, general of the Society of Jesus, to appear before him and, in front of the cardinals and prelates of the Curia, castigated him for undermining his French policy. "You are not content to blacken me in France, you also wish to tear me apart in Italy," the Pope thundered. It was an extraordinary humiliation for the leader of the Jesuits, and a public repudiation of the order as a whole. Only four years after the canonization of St. Ignatius, the once-invincible Jesuits were relegated to the farthest margins of the Roman Curia.

While the Jesuits suffered, their enemies prospered. Galileo, who was in Rome to oversee the publication of *The Assayer*, met several times with the Pope in the spring of 1624 for friendly discussions on natural philosophy. He returned to Florence in June bearing a letter that declared him the Pope's "beloved son," and with warm encouragement for his next book—which he then called "Treatise on the Flux and Reflux of the Sea," but which would ultimately become the *Dialogue on the Two Chief World Systems*. Galileo even believed that he had been given implicit permission to reopen the question of the motion of the Earth, a misjudgment that would cost him his liberty nine years later. In the autumn of 1628 the freethinking attitudes of Galileo and his friends even reached deep into the very center of Jesuit power: in a splendid ceremony in the great hall of the Collegio Romano, with numerous cardinals in attendance, the Marchese (Marquis) Pietro Sforza Pallavicino (1607–67), godson of the Cardinal of Savoy, defended his doctoral dissertation in theology. The young aristocrat was a rising star in Roman intellectual circles, with a bright future that would ultimately lead him to the college of cardinals. Even at the age of twenty-one he was already a member of the sparkling literary "Academy of the Desirous," and a friend to Galileo. On this occasion, he proved himself as one: his dissertation was a defense of the orthodoxy of the doctrine of atomism, which Galileo advocated in *The Assayer* and which was the target of Jesuit charges of unorthodoxy and even heresy. Contrary to the claims of Galileo's nemesis Father Orazio Grassi of the Collegio Romano, atomism, according to Pallavicino, was unobjectionable and perfectly consistent with the official doctrine of the Eucharist. Only a few months later, the Jesuit-trained Pallavicino became a full-fledged member of the Lincean Academy.

With their fortunes at a historically low ebb, and their authority and prestige under attack, the Jesuits effectively suspended their campaign against the infinitely small. After all, at a time when a defense of despised atomism could be publicly recited in their own most hallowed halls, how could the Jesuits credibly condemn the related doctrine of mathematical indivisibles? So they lay low for seventeen years, issuing not a single condemnation while the mathematics of the infinitely small gained ground. It was during those years that Cavalieri developed his method of indivisibles and ultimately secured the prestigious chair of mathematics in Bologna. It was also during those years that Torricelli was first introduced to the new mathematics by his mentor, Benedetto Castelli, and embarked on the career that would make him the most influential practitioner of the new mathematics. All the while, the Jesuits observed, took notes, and patiently bided their time.

THE CRISIS OF URBAN VIII

On September 17, 1631, the Protestant armies of Sweden and Saxony met the Catholic forces of the Holy Roman Empire in battle near the village of Breitenfeld in Saxony. As the Imperials charged, the inexperienced Saxons panicked and fled the field, exposing the flank of their Swedish allies to a potentially devastating strike. But the Swedes held their ground: they covered their flank and then launched their own attack. Under the cool direction of their king, Gustavus Adolphus, they routed the emperor's men, inflicting thousands of casualties on Count Tilly's previously invincible army. At a stroke, the road to the heart of Catholic Germany lay open before the Protestant Swedes.

The Swedish victory at Breitenfeld stunned Europe, instantly transforming a war that had already lasted for thirteen years and was to continue for seventeen more. Up to that point the Habsburg Imperial armies had outclassed and defeated all their rivals. They had crushed the Bohemian nobles in the Battle of White Mountain in 1620 and had defeated the Danes, who had intervened in support of their Protestant coreligionists. The union of Protestant princes, under the leadership of Johann Georg of Saxony, proved no match for the Imperial generals, Count Tilly and Albrecht von Wallenstein. But in 1630, Gustavus Adolphus of Sweden ended his war with Poland and landed his battle-tested army in northern Germany. The Holy Roman Emperor, Ferdinand II, hoped to keep the Swedes weak and isolated, but his expectations were dashed in early 1631, when Gustavus came to terms with Cardinal Richelieu of France. Cardinal though he was, Richelieu was more intent on frustrating Habsburg designs for European domination than on promoting the interests of his mother Church, so he promised to finance Gustavus's campaign on a grand scale. The results of this unorthodox alliance became clear on the field of Breitenfeld, where Gustavus's veterans —well armed, well trained, and united behind the inspired leadership of the king— crushed the empire's most powerful army.

For more than a year after the battle, the Swedes swept through Germany like a force of nature. They defeated the Habsburg armies again in April 1632, at the River Lech, killing Count Tilly and marching south into Bavaria. Gustavus's army was now deep in the heart of German Catholicism, and was busy ransacking its cities and desecrating its churches. The Jesuit colleges, proud symbols of a resurgent Catholicism, were favorite targets of the Swedes. They were plundered mercilessly, their books and treasures dispersed, the learned fathers chased away. Meanwhile, Johann Georg of Saxony, taking advantage of the Swedes' domination, invaded Bohemia, occupying and looting the former imperial

capital of Prague and disbanding the city's famed Jesuit college. That
November, the imperial army, now commanded by the crafty von Wallenstein, took a stand once more, near the town of Lützen, but the
Swedish veterans prevailed once again. Only Gustavus's death in that
battle finally slowed down the Swedish juggernaut, bringing relief to
Catholics. When the news spread across Catholic Europe, church bells
rang from Vienna to Rome, and the sons and daughters of the Church
gathered in special celebratory Masses to thank the Lord for their deliv-
erance from a ruthless enemy.

The sudden crisis in Catholic fortunes in Germany hit the papal
Curia like thunder on a clear day. At a stroke, the Pope's policy of playing France against the Habsburgs in order to preserve his own freedom
of action became untenable. It was one thing to make overtures to Richelieu when the Catholic empire reigned supreme, and with the Protestants seemingly on the run. It was quite another to do so when Richelieu
had allied himself with the heretics and the fate of Catholic Germany
hung in the balance. Urban still wavered, but he could not do so for long:
when he appeared unwilling to ally himself with the Habsburgs and
throw everything into a life-and-death struggle against the marauding
Swedes, there were those in Rome ready to remind him of his duty.

Chief among them was Cardinal Gaspar Borgia (1580–1645), the
Spanish ambassador to the Holy See and the leader of the opposition in
the Curia to the Pope's pro-French policy. Just as important, he was the
grandson of Francis Borgia, the Duke of Gandia, who had been a devoted follower of Ignatius of Loyola and ultimately the third superior
general of the Society of Jesus. The bonds between the Borgia clan and
the Jesuits remained strong, and Gaspar had been a natural ally of the
Society in Rome's culture wars. The cardinal and the Jesuits had stood
shoulder to shoulder in the 1620s in a struggle against what they saw as
Urban's tolerance of dangerous and heretical opinions, first and foremost those of Galileo and his friends. Politically, the cardinal and the
Jesuits were staunch advocates of a Habsburg alliance, which would
unite the empire, Spain, and the Papacy in a holy war against the Protestant schismatics.

Marginalized since the election of Urban VIII in 1623, Borgia and
his allies believed that the crisis of Catholic Germany had given them
the opening they needed. On March 8, 1632, in the Hall of the Consis-

tory in the Vatican, in the presence of the Pope and the cardinals, Borgia launched his attack. Breaking all rules of protocol and decorum, he startled Urban by reading an open letter harshly criticizing the Pope's policies. He denounced the unholy French-Swedish alliance and demanded that Christ's vicar on earth make his apostolic voice heard like the trumpet of redemption, uniting all Catholics in a titanic struggle against the heretics. Scandalized by the insult to the Pope, Cardinal Antonio Barberini (who was also Urban's brother) lunged at Borgia, but was repelled by the coterie of pro-Spanish and pro-Habsburg cardinals surrounding the Spanish envoy. Borgia finished reading the letter.

For a Pope to be lectured to by a mere cardinal, and be accused of letting the enemies of the faith have their way, was an insufferable humiliation. In the months that followed, Urban VIII tried to preserve his dignity and authority by punishing some of the prelates who had turned against him in his own house, and sending out indignant letters of protest to Madrid. But the game was up, and the Pope knew it. With the changing fortunes of the war, and the changing balance of political power in Rome, Urban VIII had run out of options. He backed away from his informal alliance with Richelieu and sided openly with the Habsburgs in their battle to save Bavaria and Bohemia from the Swedes. The Pope's dramatic course reversal was felt just as strongly in Rome, where—according to the Florentine ambassador Francesco Niccolini— surveillance of orthodoxy and vigilance against heretics and innovators were the instruments of power of the newly ascendant Spanish party. Urban now distanced himself from the freethinking ways of the Linceans, effectively removing his protection of Galileo. Cardinal Borgia was now the most powerful man in the Eternal City, second only to the Pope—if that. And the Jesuits, who in the nine years since Urban's accession had been exiled from the power centers of the Curia, now came in from the cold.

In 1632 the Jesuits were like a political party returning to power after a long exile, determined to put their stamp on the cultural and political life of Rome and all Catholic lands. Their first target was the man who had humiliated them for years with his sharp pen and venomous taunts, and happily for them, he provided them with a perfect opportunity to strike. In May of 1632, Galileo published his *Dialogue on the Two Chief World Systems*, the final incarnation of his book on the causes of the tides, which he had discussed with Urban years before. The *Dialogue* may not have violated the letter of Bellarmine's edict of 1616 against advocating Copernicanism, since at book's end it defers to the authority of the Church on the question of the Earth's motion. But it did, as anyone reading the book could see, violate the spirit of the edict, by eloquently presenting all the arguments in favor of Copernicanism while undercutting and ridiculing the counterarguments. Galileo, however, was apparently unconcerned, confident in the protection of the Pope and the Pope's Lincean cardinal nephew.

But Galileo's timing could not have been worse. In place of the Rome he had known, in which his friends the Linceans moved in the highest circles while the Jesuit fathers fumed helplessly on the sidelines, was a new Rome in which the positions were reversed: the Jesuits were ascendant, and Galileo's friends were running for cover. The *Dialogue* had been cleared for publication by official Vatican censors, but Galileo's enemies argued successfully that the license had been obtained under false pretenses. Galileo was charged with holding Copernicanism to be true and advocating it, thus violating Church doctrine as well as the personal injunction handed him by Cardinal Bellarmine sixteen years before. He was interrogated three times by the Inquisition, in 1632 and 1633, the last time under threat of torture. He was found to be "vehemently suspect of heresy" and sentenced to imprisonment at the pleasure of the Inquisition. After he publicly recanted his opinions, his sentence was commuted to permanent house arrest, and he spent the remainder of his life confined to his villa in Arcetri.

Rivers of ink have been spilt debating the causes of Galileo's persecution, with suggestions ranging from an irrepressible conflict between science and religion to the prickliness of a thin-skinned Urban VIII, who allegedly thought that Galileo had ridiculed him in the *Dialogue* and was out for revenge. It is a debate that has been ongoing for nearly four centuries and will undoubtedly continue for many more, and any of the causes discussed might have played some role in the events. Nevertheless, it cannot be a coincidence that the tragic reversal of Galileo's fortunes coincided precisely with the political crisis in Rome and the return to favor of his Jesuit enemies. His downfall was brought about in large measure by the enmity of the resurgent Society of Jesus.

At the very same time that the Jesuits were gearing up to indict and try Galileo for his Copernicanism, they relaunched another fight, not as publicly visible but just as crucial to their plans for reshaping the religious and political landscape: the war against the infinitely small. Relaunched in 1632, the campaign against this dangerous idea would be pursued with fierce and dogged determination for the next several decades. It would not end until infinitesimals were effectively extinguished in Italy, and significantly curtailed in other Catholic countries.

THE CENSORS, PART II

And so it was that on a Roman summer day, August 10, 1632, the Revisors General met at the Collegio Romano to pass judgment on the doctrine of the infinitely small. As was always the case, the proposition before the Revisors had allegedly been "sent in" for them to rule on, but unusually, the record of the meeting does not specify which province asked for the judgment. It is more than likely that, in this case, the initiative for the ruling came not from some anxious provincial but from the Jesuit hierarchy in Rome. Indeed, for the Jesuit leaders, the matter was urgent, as the insidious doctrine appeared to be seeping into the heart of the order itself.

Only a few months earlier, Father Rodrigo de Arriaga of Prague published his *Cursus philosophicus*, a textbook on the essential philosophical doctrines to be taught in the Society's colleges. Arriaga was not just any Jesuit father. Sent to Prague after the Imperial forces dislodged the Protestants in 1620, he had played a leading role in securing Jesuit control of Bohemian schools and universities. He soon became dean of the faculty of arts at Prague University, and was now rector of the Jesuit college there. So great was his intellectual authority and his fame as a teacher that he was the subject of a popular quip: "Pragam videre, Arriagam audire"—"To see Prague, to hear Arriaga."

Arriaga's *Cursus philosophicus* was an immediate hit among Jesuit scholars and educators, as well as many others outside the order. On the surface it seemed like a traditional, even old-fashioned text, modeled on medieval commentaries on Aristotle. Indeed, Arriaga was famous for reviving the old Scholastic disputation format, in which questions are posed on a canonical text and then discussed and answered. But to the shock of Arriaga's superiors, this seemingly sedate book expounded some very radical opinions. An entire section was devoted to a discussion of the composition of the continuum, and the conclusions reached were not at all what the Jesuit authorities would find acceptable. After carefully weighing all the arguments for and against the doctrine of indivisibles, and discussing them at length, Arriaga came to a conclusion: it is more probable than not that the continuum is indeed composed of separate indivisibles.

We do not know why Arriaga ventured into these troubled waters. Quite possibly he was influenced by his friend Gregory St. Vincent. The two taught together in Prague, and when, after the Battle of Breitenfeld, they were forced to flee the city before the advancing Saxons, it was Arriaga who saved St. Vincent's manuscripts from being dispersed and lost. Perhaps Arriaga was impressed by St. Vincent's use of infinitesimals and wanted to provide a philosophical justification for his friend's controversial mathematics. Perhaps Arriaga also thought that his stature in the order meant he was immune to censorship, and he may well have been right, had the Jesuits remained in the political wilderness they had occupied for most of the previous decade. But in 1632 the Jesuits in Rome were back in charge, and determined to end the toleration of heretics and to enforce their strict theological dogma. In this new Rome, Arriaga's challenge to his superiors could not go unanswered.

The document recording the Revisors' meeting that day, signed by Father Bidermann and his four colleagues, is preserved in the archives of the Society of Jesus in Rome. As befits the gravity of the issue at hand, the text is unusually formal. It takes up a complete page in the Revisors' records, instead of the usual line or two, and it begins with an imposing legalistic header: "Judgment on the Composition of the Continuum by Indivisibles." With a barely disguised reference to Arriaga, it continues:

> This has been handed to us to examine, proposed by a certain Professor of Philosophy, on the composition of the continuum. The permanent continuum can be constituted of only physical indivisibles or atomic corpuscles having mathematical parts identified with them. Therefore the said corpuscles can be actually distinguished from each other.

The medieval technical terminology used by the Revisors is rather opaque, but the meaning of the proposed doctrine is nevertheless clear: any continuous magnitude, physical or mathematical, is composed of irreducible, indivisible parts that can be identified one by one. What was before the Revisors was the doctrine of the infinitely small, the basis of the infinitesimal mathematics of Galileo, Cavalieri, and Torricelli.

The Revisors' judgment was unsparing:

> We judge this proposition to be not only contrary to the common doctrine of Aristotle, but that it is also improbable, and in our Society it has always been condemned and prohibited. It cannot in the future be seen by our professors as permitted.

And that was that. The doctrine of the infinitely small was now banned by the Society of Jesus for all eternity. Neither philosophers such as Arriaga nor mathematicians such as St. Vincent could henceforth be allowed to promote their subversive theories from within the Society. The order's position was now set, and any member Jesuit who challenged this dogma would face censure from the highest ranks of the order.

Once the Revisors pronounced their verdict, notices of their decision were immediately sent out to every Jesuit institution around the world. The process was routine, and usually the responsibility of the scribes and secretaries who managed the ordinary correspondence between Rome and the provinces. But when it came to the infinitely small, this was not deemed sufficient: the command to cease and desist from teaching, holding, or even entertaining the doctrine came directly from the superior general of the Jesuit order. Only a few years before, General Mutio Vitelleschi had been publicly humiliated by the Pope following the Santarelli affair. Now, with the Society poised to reshape the Catholic intellectual landscape, he was determined that, on matters of consequence, the order speak with a single, all-powerful voice.

Vitelleschi moved quickly. Only six months after the Revisors issued their edict, he found himself writing to Father Ignace Cappon at the Jesuit College in Dôle, in eastern France, to complain that his repeated instructions were not being followed. "As regards the opinion on quantity made up of indivisibles," he wrote impatiently, "I have already written to the provinces many times that it is in no way approved by me and up to now I have allowed nobody to propose it or defend it." Indeed, he had made every effort to suppress the doctrine: "If it has ever been explained or defended, it was done without my knowledge. Rather, I demonstrated clearly to Cardinal Giovanni de Lugo himself that I did not wish our members to treat or disseminate that opinion." The fight to wipe out any vestiges of the offending doctrine from the Society was one the general would undertake himself.

The lines were now firmly drawn. On the one side were the Jesuits, determined to wipe out the doctrine of the infinitely small. On the other side was a small band of mathematicians who still saw Galileo as their undisputed leader, despite his public humiliation. And the struggle raged. In 1635, Cavalieri published his *Geometria indivisibilibus*, giving the most systematic exposition of the method of indivisibles. Three years later, Galileo published his *Discourses* in Holland, which included his treatment of the paradox of Aristotle's wheel and his discussion of infinitesimals. Galileo's towering fame ensured that his views on the continuum would be taken very seriously by scholars across Europe, and his praise of Cavalieri established the Jesuat monk as the leading authority on the subject of indivisibles. In response, the Jesuits struck back with repeated condemnations of the infinitely small. From this time onward a steady stream of denunciations of the offending doctrine began issuing from the pens of the Revisors General in Rome.

On February 3, 1640, for example, the Revisors were asked to comment on the proposition that "The successive continuum . . . is composed of separate indivisibles," and ruled that "the doctrine is prohibited in the Society." Less than a year later, in January 1641, they were once again confronted with ideas "that were invented or innovated by certain moderns," including the proposition "that the continuum is made of indivisibles," and a variation that claimed "that the continuum is made of indivisibles that expand and contract." Both propositions, the Revisors ruled, were "contrary to the common and solid opinion." On May 12, 1643, they came down hard on an author who allegedly preferred the opinions of Zeno to those of Aristotle: "We do not approve or concede of these," they wrote, "as it is against the Society's constitution and rules, as well as the decrees of the General Congregation."

A pattern set in. In Rome the Jesuits were strong enough to quash any talk of the forbidden doctrine; but farther away, some mathematicians could still defend and promote it. Torricelli, for example, was in Rome in the 1630s, toiling in private on the new mathematics but publishing not a word. But it took only two years after his installment as Galileo's successor at the Medici court in Florence before he published his *Opera geometrica*, which contains the fruits of his silent labor in Rome. Even from the safety of his court position, however, Torricelli tried to avoid open conflict with his powerful critics. Unlike Galileo, he did not directly engage with them in his book, argue over the merits of his method, or ridicule their motives or reasoning. He let his powerful results speak for themselves, and that they undoubtedly did: the *Opera geometrica* was admired and emulated by mathematicians from Germany to England. In Italy the long shadow cast by the Society of Jesus was sufficient to ensure that no mathematician there expressed similar enthusiasm for Torricelli's work. But the Jesuits nonetheless took note of his success, and prepared to strike back.

Three years later, Cavalieri fired his last salvo in the fight over the infinitely small. As professor of mathematics at the University of Bologna, he enjoyed the protection of his own order, the Jesuats of St. Jerome, as well as of the Bolognese Senate, to which he dedicated several of his books. Like Torricelli, Cavalieri, too, was far from Rome, and in a position in which he could afford to risk the wrath of the domineering Jesuits. In 1647 he was in any case mortally ill, and did not need to concern himself with any future retaliation by his enemies. Shortly before his death, he managed to see through the publication of his second, and final, book on indivisibles, the *Exercitationes geometricae sex*. As regards to mathematical method, the *Exercitationes* contained little that was new to anyone familiar with Cavalieri's monumental *Geometria indivisibilibus*. As it happened, no one except for the author saw any fundamental difference between the two. But the book did play a significant role in the ongoing struggle over infinitesimals: in a wholly new section, Cavalieri directly attacked the Jesuit mathematician Paul Guldin, who had harshly criticized Cavalieri's approach. It was the last statement on indivisibles from the man who was widely recognized as the greatest authority on the method, but it could not stem the tide of Jesuit denunciations. In 1649 the Revisors ruled on two more variations of the doctrine of infinitesimals, proposed by the Jesuits of Milan. As usual, they decreed that the doctrine was forbidden and could not be taught in the Society's schools.

THE HUMBLING OF THE MARQUIS

Arriaga's views on the continuum were unequivocally condemned by Father Bidermann and his Revisors in 1632, but the *Cursus philosophicus* prospered nonetheless. Not only did the book retain its popularity, but Arriaga managed to secure permission from his superiors to publish new editions every few years, the last one appearing two years after his death in 1667. The reason for this leniency was not, as Arriaga suggests in his introduction to the posthumous edition, that he had published his opinions in good faith and that they "do not pertain to matters of faith," as such considerations did not stop the Jesuit hierarchy from taking an extremely grim view of any advocacy of indivisibles. It was surely Arriaga's enormous popularity as a teacher and his standing as one of the leading intellectuals in Europe that allowed his unorthodox book to remain in print. Arriaga brought intellectual stature and prestige to an order eager to assert its intellectual leadership all across Europe, and the Jesuit superior generals therefore thought it best to leave him in peace. But once Arriaga died, there was no more need to accommodate such dissent. The 1669 edition, approved before Arriaga's death and published shortly after it, was to be the last of the *Cursus philosophicus*.

At least one other high-ranking Jesuit tried to emulate Arriaga's example: the Marchese Pietro Sforza Pallavicino, who as a young maverick in 1620s Rome dared to challenge the Jesuits in a public oration in their own halls. When the political tide in Rome turned against the Galileans, Pallavicino paid a price for his cockiness. In 1632 he was exiled from the papal court and sent to govern the provincial towns of Jesi, Orvieto, and Camerino. But by 1637 the marchese had tired of country living and was ready to see the light: in a reversal that left Roman society dumbfounded, Pallavicino took the monastic oaths and entered the Society of Jesus as a novice. For the Jesuits it was a stunning coup. Not only was Pallavicino a high-ranking nobleman and a renowned poet and scholar, but he was also famous for his open criticism of the Society. Nothing demonstrated the Jesuits' triumph as clearly as the defection of the brilliant marchese from their enemies' camp, to enter their own ranks as a humble novice.

Even so, Pallavicino was no ordinary novice. It took Clavius, who had come from humble stock, two decades to climb from novice to professor at the Collegio Romano. The noble Pallavicino completed the same journey in two years, before being appointed professor of philosophy—an honor Clavius never attained. It seems likely that General Vitelleschi had struck a deal with the young marchese, promising him a shortened apprenticeship and a prestigious appointment in Rome as the price of his entry into the Society. Be that as it may, by 1639, Pallavicino was teaching philosophy at the Collegio Romano, and several years later he was also appointed professor of theology, the highest academic post at the college. In 1649 he published a comprehensive defense of the Jesuits, entitled *Vindicationes Societatis Iesu*, a particularly appropriate task for one who some years before had been among the Society's most public critics. At the personal request of the Pope, he then authored a history of the Council of Trent, intended as an official rebuttal to the controversial (and, to the Papacy, defamatory) history that Venetian Paolo Sarpi published back in 1619. Pallavicino's volumes came out in 1656 and 1657 as *Istoria del Concilio di Trento*, and earned the marchese the crowning honor of his life: the cardinal's purple.

Pallavicino's career among the Jesuits went from triumph to triumph, but in the 1640s he did suffer some embarrassing setbacks. Despite throwing in his lot with Galileo's enemies, the marchese still considered himself a progressive thinker and an admirer of the Florentine master. Such residual allegiance to the Jesuits' vanquished enemy was bound to be viewed with suspicion by the Society's hierarchy, and indeed, Pallavicino frequently came under the Revisors' scrutiny for what they considered his "novel doctrines." Nevertheless, encouraged no doubt by the example of Arriaga and believing that his stature would protect him from censure, Pallavicino forged ahead, lecturing on his unorthodox views to the Collegio's students. But the marchese had miscalculated. He himself hints as much in the *Vindicationes Societatis Iesu*, when he recalls having "to face a fight several years ago" when he wished to express himself on a matter he considered "common or well known." Pallavicino, it seems, was criticized and perhaps chastised for his position, but he was not one to concede that he was in the wrong. To the contrary, he insists, although the propositions he mentioned may be false, in a Society so devoted to the well-being of its students, "a certain freedom to speak of positions less accepted should, up to a point, not be eliminated, but promoted."

Pallavicino tried to put the best face on the incident, but a much clearer picture of the events emerges in an angry letter from Superior General Vincenzo Carafa, who had succeeded Vitelleschi, to Nithard Biberus, Jesuit provincial for Lower Germany. "When I came to know there are some in the Society who follow Zeno, who pronounced in a philosophy course that a quantity is composed of mere points, I let it be known that it is not approved by me," the superior general wrote irately on March 3, 1649. "And since in Rome Father Sforza Pallavicino taught this, he was ordered to retract it in the very same course." It was a stinging rebuke, and undoubtedly a humbling experience for the marchese to be forced to retract his own words before his own students. The bitterness of the experience still echoes in the *Vindicationes*, but as a member of a society that valued obedience above all, refusing a direct order from the superior general was out of the question. So Pallavicino swallowed his pride, retracted his teachings on the infinitely small, and having learned his lesson, quietly resumed his climb up the ladder of the Jesuit hierarchy.

Carafa's letter to Biberus makes clear that the superior general could not allow even the marchese to get away with teaching the forbidden doctrine. When he recently wrote to censure a professor in Germany for teaching it, Carafa continued, the professor answered, "by way of excuse, that Arriaga and a certain Portuguese of ours have expounded these views in print." The general, however, would have none of it: "I then wrote again that these [two] works being given, there will be no third who will imitate them." Arriaga (as well as the unnamed Portuguese) was a special case, grandfathered in from laxer times. But no one, not even the Marchese Pallavicino, should consider it a precedent. The doctrine of the infinitely small was banned to all Jesuits, and anyone who dared promote it would suffer the consequences.

A PERMANENT SOLUTION

But even as the superior general was personally admonishing his subordinates and publicly humbling an excessively proud Jesuit, pressure was growing for a more permanent solution to the problem of dissent within the Society. Already in 1648, Carafa had instructed the Revisors to search their records and come up with a provisional list of theses that should be permanently banned from the order. When, following Carafa's death, a General Congregation convened in December 1649, it instructed the newly elected superior general, Francesco Piccolomini, to follow up on his predecessor's initiative. Over the next year and a half, a Jesuit committee met and devised an authoritative list of prohibited doctrines. The results of their labors were published in 1651 as part of the *Ordinatio pro studiis superioribus* ("Regulations for Higher Studies"), designed to preserve the Society's "solidity and uniformity of doctrine." Henceforth, any Jesuit anywhere in the world would have access to an authoritative list detailing which doctrines were anathema to his order and must never be held or taught.

The sixty-five forbidden "philosophical" theses cited in the *Ordinatio* (there are also twenty-five "theological" ones) make for an eclectic list. Some banned propositions infringed on accepted interpretations of Aristotelian physics, such as "primal matter can naturally be without form" (number 8), or "heaviness and lightness do not differ in species, but only in regard to more or less" (number 41). Some offending propositions were tinged with materialism, such as "the elements are not composed of matter and form, but only from atoms" (number 18). Other theses were banned for challenging divine omnipotence, such as "a creation so perfect is possible that God is incapable of creating a more perfect one" (number 29); and still others were banned for teaching of the diurnal motion of the Earth (number 35) or promoting the magical curing of wounds at a distance (number 65). But no fewer than four of the prohibited propositions directly addressed the question of the composition of the continuum from indivisible parts:

25. The succession continuum and the intensity of qualities are composed of sole indivisibles.
26. Inflatable points are given, from which the continuum is composed.

30. Infinity in multitude and magnitude can be enclosed between two unities or two points.
31. Tiny vacuums are interspersed in the continuum, few or many, large or small, depending on its rarity or density.

Thesis 25 is the broadest of the four, referring to any possible continuum and its composition. The question of the "intensity of qualities" refers to medieval debates in which the intensity of qualities such as "hot" or "cold" existed along a gradient, raising the question of whether there was a finite or infinite number of grades of these qualities forming the continuum. Thesis 26 was a response to widespread speculation in the seventeenth century about what caused a material's change in density, a question that was considered one of the toughest challenges to an atomic theory of matter. The thesis claimed that matter was composed of innumerable inflatable points, whose size at any given moment determined the degree of density or rarity, but to the Jesuits this was no more acceptable than the simpler doctrine that the continuum was composed of indivisibles. Thesis 30 is the most explicitly mathematical of the four, referring directly to the method of indivisibles practiced by Cavalieri and Torricelli, which relied on the division of finite lines or figures into an infinite number of indivisible parts. Thesis 31 appears to address Galileo's theory of the continuum as expounded in the *Discourses* of 1638. Relying on analogies with matter and the paradox of Aristotle's wheel, Galileo concluded that the continuum is interspersed with an infinity of tiny vacuums. Between them, these four propositions encompass the different variations of the method of indivisibles under dispute in the middle of the seventeenth century. All were unequivocally banned.

The *Ordinatio* of 1651 was a turning point in the Jesuit battle against the infinitely small. The prohibition on the doctrine was now permanent, and was backed by the highest authority in the order, the General Congregation. Printed, published, and widely disseminated, the *Ordinatio* was brought to the attention of every Jesuit father teaching at every one of the Society's institutions across the world. The repeated blasts by the Revisors, who every few years issued their censures of the doctrine, were now at an end. No further condemnations were needed: the prohibition was now permanent and compulsory, and every member of the order knew it well. And so it stood for the next century, as the *Ordinatio* remained the fundamental guide to Jesuit teachings. In fact, the document did much more: it set the tone for intellectual life in the lands where the Jesuits dominated. For the few and lonely mathematicians still defending infinitesimals in Italy, the consequences were devastating.

5

The Battle of the Mathematicians

GULDIN VERSUS CAVALIERI

"The reasoning by indivisibles convinces all the famous geometers brought up here," wrote Stefano degli Angeli (1623–97), professor of mathematics at the University of Padua, in 1659. Angeli, as was his way, expressed himself with great bravado, but the truth was very different. By the time he was writing, Angeli was likely the last mathematician left in Italy to adopt the method of indivisibles, and one of the even fewer to publish their work in the field. Most of the individuals named in his own list of the "famous" adherents of the method resided north of the Alps, many in France and England. The Italians on the list belonged to an old and dying generation of Galileans, who had stopped publishing on the method decades before. When he wrote those words, Angeli was not in fact reporting on a favorable state of affairs, but rallying the troops for a desperate rearguard action on behalf of infinitesimals, which were in danger of being extinguished in the land where they had once flourished. Who his enemies were, he had no doubt: "the three Jesuits, Guldin, Bettini, and Tacquet" were, he claimed, the only ones who remained unconvinced by the method of indivisibles. "By what spirit they are moved," he continued, with evident frustration, "I do not know."

Paul Guldin, Mario Bettini (1584–1657), and André Tacquet were among the most notable mathematicians of the Society of Jesus in the mid-seventeenth century. Tacquet, whom we have already met, was the most original and creative of the three, but Guldin, too, was a widely respected mathematician. Bettini was perhaps less so, known mostly as a prolific author of rambling collections of mathematical results and curiosities, and less as a creative thinker in his own right. But he, too, was known in the Society and beyond as a man of broad learning and considerable authority on things mathematical. Together, Guldin, Bettini, and Tacquet were a formidable trio, exemplifying the intellectual prowess and the cultural and political cachet of the Jesuit mathematical school. And in the 1640s and '50s, all three were engaged in the same mission: to discredit and undermine the method of indivisibles using sound and incontrovertible mathematical arguments. Theirs was a dimension of the Jesuit war on infinitesimals that was just as critical as the repeated condemnations of the Revisors and the *Ordinatio* of 1651. For if the use of infinitesimals in mathematics was to be permanently abolished, it was not enough to declare them philosophically, theologically, or even morally wrong, and legally banish them. It was also crucial to prove them mathematically wrong.

Guldin, the oldest of the three, was the first to take the field. Born Habakkuk Guldin to Protestant parents of Jewish descent in St. Gall, Switzerland, he may be the first in a long and illustrious line of Jewish (and converted Jewish) mathematicians that continues to this day. Guldin was not raised to be a scholar but an artisan, and he was working as a goldsmith when he began having doubts about his Protestant faith. At age twenty he converted to Catholicism and joined the Jesuits, changing his name in the process from the Old Testament prophet Habakkuk to Paul, the most famous Jewish convert, who preached the Christian faith to the Gentiles. Guldin's was an eclectic and unusual background for a Jesuit, encompassing as it did many of the religious and ethnic fault lines of the early modern world, but this did not prevent his full acceptance in the Society. Indeed, it is one of the most admirable characteristics of the early Jesuits that despite pressure from the Iberian kingdoms, which placed the highest value on *limpieza de sangre* (purity of blood), the Society remained one of the most welcoming Catholic institutions for converts of all kinds.

The Society of Jesus was also, to a large degree, a meritocracy, and although high-born noblemen such as Marchese Pallavicino enjoyed enviable advantages, there was also a path forward for men of humble origins such as Guldin. As a bright young man with a talent for mathematics, he rose steadily through the ranks of the order, and was ultimately sent to the Collegio Romano to study with Clavius. Guldin spent fewer than three years under Clavius's tutelage before the old master passed away in 1612. Five years later he was sent to teach mathematics in the Habsburg Austrian lands, and spent the rest of his life at the Jesuit college in Graz and at the University of Vienna. Nevertheless, it is clear from Guldin's subsequent career that his years with Clavius shaped his mathematical outlook for a lifetime. Guldin was Clavius's follower in every way: he adhered to the old Jesuit's view that mathematics lies halfway between physics and metaphysics, he believed in the primacy of geometry among mathematical disciplines, and he insisted on following the classical Euclidean standards of deductive proof. All these positions made him an ideal choice as a critic of the method of indivisibles.

Guldin's critique of Cavalieri's indivisibles is contained in the fourth book of his *De centro gravitatis* (also called *Centrobaryca*), published in 1641. He first suggests that Cavalieri's method is not in fact his own, but was derived from that of two other mathematicians: one was Johannes Kepler, who, though a Protestant, was Guldin's friend in Prague; the other, the German mathematician Bartholomew Sover. The charge of plagiarism is almost certainly unmerited, and in any case, Guldin was not paying Kepler or Sover much of a compliment, as he soon launches into a harsh and penetrating critique of the method.

Cavalieri's proofs, Guldin argues, are not constructive proofs of the kind that classically trained mathematicians would accept. This is undoubtedly true: in the Euclidean approach, geometrical figures are constructed step by step, from the simple to the complex, with the aid of only a straightedge and a compass, for the construction of lines and circles, respectively. Every step in a proof must involve such a construction, followed by a deduction of the logical implications for the resulting figure. Cavalieri, however, proceeded the other way around: he began with ready-made geometrical figures such as parabolas, spirals, and so on, and then divided them up into an infinite number of parts. Such a procedure might be called "deconstruction" rather than "construction," and its purpose is not to erect a coherent geometrical figure, but to decipher the inner structure of an existing one. Such a procedure, the classically trained Guldin is quick to point out, did not conform to the rigorous standards of a Euclidean demonstration, and should be rejected on those grounds alone.

Guldin next goes after the foundation of Cavalieri's method: the notion that a plane is composed of an infinitude of lines, or a solid of an infinitude of planes. The entire idea, Guldin argues, is nonsense: "In my opinion," he writes, "no geometer will grant him that the surface is, and could in geometrical language be called 'all the lines of such a figure.' Never in fact can several lines, or all the lines, be called surfaces; for, the multitude of lines, however great that might be, cannot compose even the smallest surface." In other words, since lines have no width, no number of them placed side by side would cover even the smallest plane. Cavalieri's attempt to calculate the area of a plane from the dimensions of "all its lines" was therefore absurd. This then leads Guldin to his final point: Cavalieri's method was based on establishing a ratio between "all the lines" of one figure and "all the lines" of another. But, Guldin insists, both sets of lines are infinite, and the ratio of one infinity to another is meaningless. No matter how many times one multiplied an infinite number of indivisibles, they would never exceed a different infinite set of indivisibles. In other words, Cavalieri's supposed ratios between "all the lines" of one figure and those of another violated the axiom of Archimedes, and were therefore invalid.

When taken as a whole, Guldin's critique of Cavalieri's method embodies the core principles of Jesuit mathematics. Clavius and his descendants in the Society all believed that mathematics must proceed systematically and deductively, from simple postulates to ever-more-complex theorems, describing universal relations between figures. Constructive proofs, moving step by logical step from lines and circles to complex constructions, are the embodiment of this ideal. Slowly but surely they build up a rigorous and hierarchical mathematical order which, as Clavius had shown, brings Euclidean geometry closer to the Jesuit ideal of certainty, hierarchy, and order than any other science. Guldin's insistence on constructive proofs was consequently not a matter of pedantry or narrow-mindedness, as Cavalieri and his friends thought: it was an expression of the deeply held convictions of his order.

The same was true of Guldin's criticism of the division of planes and solids into "all the lines" and "all the planes." Mathematics must be not only hierarchical and constructive, but also perfectly rational and free of contradiction. But Cavalieri's indivisibles, as Guldin points out, were incoherent at their very core, since the notion that the continuum is composed of indivisibles simply does not stand the test of reason. "Things that do not exist, nor could they exist, cannot be compared," he asserts with impeccable reasoning. It is therefore no wonder that they lead to paradoxes and contradiction, and ultimately to error. To the Jesuits, such mathematics was far worse than no mathematics at all. If this flawed system were accepted, mathematics could no longer be the basis of an eternal, rational order. The Jesuit dream of a strict universal hierarchy as unchallengeable as the truths of geometry would be doomed.

In his writings, Guldin does not explain the deeper philosophical reasons for his rejection of indivisibles; nor do Bettini and Tacquet. At one point Guldin comes close to admitting that there are greater issues at stake than the strictly mathematical ones, writing cryptically that "I do not think that the method [of indivisibles] should be rejected for reasons that must be suppressed by never inopportune silence," but he gives no explanation of what those "reasons that must be suppressed" could be. The three Jesuits' reticence to disclose any nonmathematical motivations for their stances is, however, quite natural. As mathematicians, they had the job of attacking the indivisibles on strictly mathematical, not philosophical or religious, grounds. Their authority and credibility would only have suffered if they had announced that they were moved by theological or philosophical considerations.

Those involved in the fight over indivisibles knew, of course, what was truly at stake. When Angeli wrote facetiously that he did not know "what spirit" moved the Jesuit mathematicians, and when Guldin hinted at "reasons that must be suppressed," they were referring to the Jesuits' ideological opposition to infinitesimals. Nevertheless, with very few exceptions, these broader considerations were never openly acknowledged in the mathematical debate. It remained a technical controversy between highly trained professionals on which procedures are allowable in mathematics and which are not. When Cavalieri first encountered Guldin's criticism in 1642, he immediately began work on a detailed refutation. Initially he intended to respond in the form of a dialogue between friends, of the type favored by his mentor, Galileo. But when he showed a short draft to Giannantonio Rocca, a friend and fellow mathematician, Rocca counseled against it. In order to avoid Galileo's fate, Rocca warned, it was safer to stay away from the inflammatory dialogue format, with its witticisms and one-upmanship that were likely to enrage powerful opponents. Much better, Rocca advised, to write a straightforward response to Guldin's charges, focusing on strictly mathematical issues, and to refrain from Galilean provocations. What Rocca left unsaid was that Cavalieri, in all his writings, showed not a trace of Galileo's genius as a writer, nor of his ability to present complex issues in a witty and entertaining manner. It is probably for the best that Cavalieri took his friend's advice, sparing us a "dialogue" in his signature ponderous and near-indecipherable prose. Instead, Cavalieri's response to Guldin is included as the third "exercise" of the *Exercitationes*, and is titled, plainly enough, "In Guldinum" ("On Guldin").

In his response, Cavalieri does not appear overly troubled by Guldin's critique, quickly dismissing the charges of plagiarism before moving on to the mathematical issues. He denies that he posits that the continuum is composed of an infinite number of indivisible parts, arguing that his method does not depend on this assumption. If one believes that the continuum is composed of indivisibles, then, yes, "all the lines" together do indeed add up to a surface, and "all the planes" to a volume; but if one does not accept that the lines comprise a surface, then there is undoubtedly something there in addition to the lines that makes up the surface, and something in addition to the planes that makes up the volume. None of this, he argues, has any bearing on the method of indivisibles, which compares "all the lines" or "all the planes" of one figure with those of another, regardless of whether they actually compose the figure.

Cavalieri's argument here may be technically acceptable, but it is also disingenuous. Anyone reading the *Geometria* or the *Exercitationes* can have no doubt that they are based on the fundamental intuition that the continuum is indeed composed of indivisibles. Cavalieri's coyness notwithstanding, the only possible reason for comparing "all the lines" and "all the planes" of figures is the belief that they in some way make up the surface and volume of the figures. The very name that Cavalieri gave his approach, the method of indivisibles, says as much, and his famous metaphors of the cloth and the book make it crystal clear. Guldin is perfectly correct to hold Cavalieri to account for his views on the continuum, and the Jesuat's defense seems like a rather thin excuse.

Cavalieri's response to Guldin's insistence that "an infinite has no proportion or ratio to another infinite" is hardly more persuasive. He distinguishes between two types of infinity, claiming that "absolute infinity" indeed has no ratio to another "absolute infinity," but "all the lines" and "all the planes" have not an absolute but a "relative infinity." This type of infinity, he then argues, can and does have a ratio to another "relative infinity." As before, Cavalieri seems to be defending his method on abstruse technical grounds, which may or may not be acceptable to fellow mathematicians. Either way, his argument bears no relation to the true reasoning or motivation behind the method of indivisibles.

That motivation comes to light only in Cavalieri's response to Guldin's charge that he does not properly "construct" his figures. Here Cavalieri's patience is at an end, and he lets his true colors show. Guldin claimed that every figure, angle, and line in a geometrical proof must be carefully constructed from first principles; Cavalieri flatly denies this. "For a proof to be true," he writes, "it is not necessary to describe actually these analogous figures, but it is sufficient to assume that they have been described mentally . . . and in consequence nothing contradictory may be deduced if we assume that these figures have been already constructed."

Here, finally, is the true difference between Guldin and Cavalieri, between the Jesuits and the indivisiblists. For the Jesuits, the purpose of mathematics was to establish the world as a fixed and eternally unchanging place, in which order and hierarchy could never be challenged. That is why each item in the world must be carefully and rationally constructed, and why any hint of contradictions or paradoxes could never be allowed to stand. It was a "top-down" mathematics, whose purpose was to bring rationality and order to an otherwise chaotic world. For Cavalieri and his fellow indivisiblists, it was the exact reverse: mathematics began with a material intuition of the world—that plane figures were made up of lines and volumes of planes, just as a cloth was woven of thread and a book compiled of pages. One does not need to rationally construct such figures, because we all know they already exist in the world. All that is needed, as Cavalieri says, is to assume and imagine them, and then proceed to investigate their inner structure. Ultimately, he continues, "nothing contradictory can be deduced," because the fact that the figures exist guarantees that they are internally consistent. If we encounter seeming paradoxes and contradictions, they are bound to be superficial, resulting from our limited understanding, and can either be explained away or used as a tool of investigation. But they should never stop us from investigating the inner structure of geometrical figures, and the hidden relationships between them.

For classical mathematicians such as Guldin, the notion that you could base mathematics on a vague and paradoxical intuition of matter was absurd: "who will be the judge" of the truth of a geometrical construction, he mockingly asks Cavalieri, "the hand, the eye, or the intellect?" But the charge of practicing an irrational geometry of the hand or eye did little to dissuade Cavalieri, since his method was indeed based on such practical intuitions. To him, Guldin's insistence that the method must be abandoned because of apparent contradictions was pointless pedantry, since everyone knew that the figures did exist, and it made no sense to argue that they shouldn't. Such nitpicking, it seemed to Cavalieri, could have grave consequences: if Guldin prevailed, a powerful method would be lost, and mathematics would be betrayed.

BETTINI'S STING

By the time Cavalieri published his response to Guldin, the Jesuit had already been dead for three years, and he himself had only months to live. But the death of the two chief protagonists, along with the passing of Torricelli in 1647, did nothing to tamp down the debate. Mathematicians could come and go, but the Society's determination to extinguish the infinitely small remained the same, and the role of chief critic of indivisibles was simply handed on to another Jesuit mathematician. Mario Bettini, who inherited the mantle from Guldin, did not claim to be a leading mathematical light; nor was he considered one by his contemporaries. His claim to fame was as the author of two very long and eclectic books of mathematical curiosities, which he called the *Apiaria universae philosophiae mathematicae* ("Beehive of Universal Mathematical Philosophy"), published in 1642, and the *Aerarium philosophiae mathematicae* ("Treasury of Mathematical Philosophy"), published in 1648. Both of these were fine exemplars of the Jesuit approach to mathematics, highlighting as they did the ways in which geometrical principles pervaded the world. They included mathematical discussions of the flight of projectiles, construction of fortifications, the art of navigation, and so on, all governed by the universal and unassailable principles of geometry. The theory of indivisibles was not a natural fit in these practically oriented and eclectic collections, but it was nevertheless the focus of book 5 of volume 3 of the *Aerarium*. This was, after all, one year after the publication of Cavalieri's rebuttal of Guldin, and it was imperative that the Society respond and keep up the pressure on the champions of the infinitely small.

It is very likely that Bettini and Cavalieri knew each other personally, and there is much to suggest that their relationship was far from friendly. In 1626, Cavalieri was appointed prior of the Jesuat house in Parma, where Bettini was professor at the university, and it is hard to imagine that the two mathematicians did not cross paths in this modest-size city. Cavalieri, it will be recalled, entertained hopes of being appointed to a mathematical chair at the University of Parma, but, as he complained to Galileo on August 7 of that year, it all came to naught. "As for the lectureship in mathematics," he wrote, "were the Jesuit fathers not here I would have great hope, because of the great inclination of Monsignor Cardinal Aldobrandini to favor me . . . but as [the university] is under the rule of the Jesuit Fathers, I cannot hope any longer." There can be little doubt that among the Jesuit Fathers who scuttled Cavalieri's appointment was their own leading mathematician, Mario Bettini.

Cavalieri did have a measure of revenge when, in 1629, he became professor of mathematics in nearby Bologna. The appointment of a Galilean to this prestigious chair in the most ancient university in Europe was a stinging blow to the Jesuits, but particularly to Bettini, who was himself a native of Bologna. In the following years, perhaps in response to Cavalieri's appointment or perhaps out of concern about their limited influence in that city, the Jesuits made plans to transfer their entire faculty from Parma to a new Jesuit college in Bologna. The move was ultimately blocked by the city's senate, which in 1641 passed an ordinance forbidding the teaching of university subjects by anyone not on the university rolls. It is easy to see Bettini and Cavalieri lining up on opposite sides of this fight: the Jesuit, eager to establish a beachhead for his order in his native city, pushing hard for the new college; the Jesuat, burned by his experience in Parma and grateful to the Bolognese Senate for allowing him to pursue his mathematical work in peace, doing everything in his power to fend off the Jesuit invasion.

What Bettini lacked in mathematical sophistication he made up for in fervency. Guldin, and later Tacquet, kept the debate largely within the bounds of technical mathematics, but Bettini did not hesitate to use blunt language and warn darkly of dire consequences if his admonitions were not heeded. It is possible that the bitterness of his personal history with Cavalieri led him to go beyond his mandate of a sedate mathematical critique of indivisibles, but whatever his personal motivation, Bettini's attitude was probably more in line with the true tenor of the Jesuit campaign against the infinitely small. He merely gave it voice.

Mathematically, Bettini added nothing of substance to Guldin's critique, but hammered away at one point alone: that "infinity to infinity has no proportion," and it therefore made no sense to compare the infinite lines of one figure with the infinite lines of another. Since this procedure is at the heart of the method of indivisibles, Bettini insisted that it was imperative that students and novices be warned against this tempting but false approach. "In order to set forth the elements of geometry," he writes, "I point out [these] hallucinations, so that novices will learn to distinguish (as in the proverb) 'what separates the false coin from the true' in geometrical philosophy." Indivisibles, according to Bettini, were a dangerous fantasy that was best ignored if at all possible. Under the circumstances, however, "being pressed, I respond to the counterfeit philosophizing about geometrical figures by indivisibles. Far, far be it from me to wish to make my geometrical theorems useless, lacking demonstrations of truth. Which would be to compare . . . figures and philosophize about them by indivisibles." To avoid undermining all demonstrations and subverting geometry itself, one must steer clear of the dangerous hallucination—the method of indivisibles.

THE COURTLY FLEMING

In 1651 André Tacquet, the urbane Fleming whose work was celebrated by Catholics and Protestants alike, published his *Cylindricorum et annularium libri IV* ("Four Books on Cylinders and Rings"), a work dedicated

Frontispiece of Tacquet's *Cylindricorum et annularium libri IV*.
(Photograph courtesy of the Huntington Library)

to the study of the geometrical features of these figures and their applications. Befitting a Jesuit publication, the frontispiece shows two angels, bathed in divine light, holding up a ring enclosing the book's title. On the ground below them, a band of cherubs is busy putting the theory into practice. The implication is clear: divine mathematics, universal and perfectly rational, orders and arranges the physical world to the best possible effect. It is a fetching visual depiction of the Jesuit view of the role and nature of mathematics.

The *Cylindricorum et annularium* is Tacquet's most celebrated work, the one that established his reputation as one of Europe's most original and creative mathematicians. As it turned out, it may have been a bit too "original and creative" for his superiors: when Tacquet sent a copy of the book to the newly appointed superior general, Goswin Nickel, the general's response was surprisingly cool. After thanking the mathematician and congratulating him on the book, Nickel added that it would be better if Tacquet applied his impressive gifts to producing textbooks of elementary geometry for use by students at the Society's colleges, rather than original works aimed at a select audience of professional mathematicians. Nickel wrote this not out of hostility to Tacquet, but as a way of giving voice to the Jesuits' suspicion of novelty, and their belief that the role of mathematics was to establish a fixed, unchanging order. Quite likely he was also uncomfortable with the fact that Tacquet made use of indivisibles in his work, if only as a means of discovery, not of proof. In any case, Tacquet, a good soldier in the Army of Christ, obeyed. From then on he published no more original work, but concentrated instead on producing textbooks, some of which were of such quality that they became standards in the field for over a century.

Born in 1612, Tacquet was far younger than Guldin or Bettini, and so had witnessed the spread of infinitesimals beyond the Alps, with Roberval in France and Wallis in England leading the charge. Bettini, with no international reputation to protect, could afford to bluntly denounce indivisibles as "hallucinations." But Tacquet could not be so disrespectful of an approach that was gaining favor among his colleagues in northern Europe. So it may have been the pressure of circumstances, it may have been the fact that, unlike Bettini, Tacquet bore no personal grudge against Cavalieri and his followers, or it may simply have been a matter of personal temperament, but when Tacquet turned his attention to the question of indivisibles, his tone was far more restrained than Bettini's shrill denunciations.

In his critique, Tacquet is respectful, even deferential, toward his rivals. He refers to Cavalieri as "a noble geometer," and insists that he "does not wish to detract from the deserved glory" of Cavalieri's "most beautiful invention." Tacquet knew of what he spoke, because he was himself deeply familiar with the work of Cavalieri and Torricelli, and was no less capable than they of using their method to arrive at new results. But once one gets beyond his congenial style and mathematical mastery, it becomes clear that Tacquet's opposition to the infinitely small is just as unyielding as that of the bruising Bettini. "I cannot consider the method of proof by indivisibles as either legitimate or geometrical," he states flatly at the opening of his discussion of indivisibles. "It proceeds from lines to surfaces, from surfaces to solids, and applies to the surface the equality or proportion obtained from the lines, and transfers what was obtained from the surfaces to the solid." "By this method," he concludes, "nothing can be proven by anyone."

Unlike Guldin or Bettini, Tacquet did not think that the method of indivisibles was completely useless. It was, for him, a practical tool for finding out new geometrical relations and testing them. But one should never mistake a result arrived at by indivisibles for a properly proven geometrical truth. "If a theorem is proposed, that is proven by no other method than indivisibles, I will always doubt its truth until it is shown that the proof can be redone through homogenes" (*homogenes* being Tacquet's term for classical-type demonstrations). Reasoning by indivisibles, he points out, though sometimes useful, is just as likely to lead to erroneous and absurd results as it is to true ones, and therefore must never be trusted.

In the substance of his critique, Tacquet follows closely in the footsteps of his predecessors Guldin and Bettini. He gladly concedes that a line can be formed by moving indivisibles—a line by a moving point, a surface by a moving line, a solid by a moving surface. But this does not mean that a quantity is composed of indivisibles, since accepting such an idea, he insists, would be the death of geometry. So, in the end, it was the refined and urbane Fleming André Tacquet, and not the coarse and quarrelsome Italian Mario Bettini, who provided a synopsis of the Jesuit position on indivisibles: if indivisibles were not destroyed, then geometry itself would be. There was no middle ground.

THE HIDDEN CAMPAIGN

The Jesuit campaign against the infinitely small in the seventeenth century proceeded on several parallel tracks. The legal track was mostly carried out in the decrees of the Revisors General, supported by direct orders from the superior generals, and punishments to recalcitrant subordinates. The mathematical track was carried out by the order's professional mathematicians Guldin, Bettini, and Tacquet. Their role was to discredit infinitesimals on purely mathematical grounds, while advocating for the methods of the ancients. The towering prestige of the Jesuits and the solidarity of the Society's mathematicians, who supported one another's positions, ensured that both the official decrees and the mathematical opinions would resonate far beyond the confines of the order.

There is much, nonetheless, that we do not know about the decades-long campaign against the infinitely small: How many mathematicians who privately supported the method of indivisibles chose to keep quiet for fear of Jesuit reprisals? How many were denied university appointments because of their suspected allegiance to the forbidden doctrine? How many aspiring mathematicians simply turned away from the infinitely small, fearing that their professional prospects would suffer if they supported it? This hidden facet of the Jesuit campaign, conducted through personal interaction, private correspondence, and institutional pressure, is very hard to pin down with any certainty. But we do know enough to get a taste of the hostility and pressure faced by mathematicians who supported the infinitely small in Italy, the land where the power of the Jesuits was greatest.

The Society's influence in Italy was deep and pervasive. Even in the 1620s, when Jesuit power was at a low ebb, Cavalieri tried for years to secure for himself a university chair, before finally being appointed professor at the University of Bologna in 1629. At least one of these rebuffs, at Parma in 1626, was due to stiff Jesuit opposition. Torricelli developed his mathematics privately in the 1630s but was never a serious candidate for any university post, and he published his work only after being installed in the Medici court in Florence. It seems more than likely that the hidden hand of the resurgent Jesuits was at work here, reaching out to extinguish any opportunity the brilliant young mathematician had to establish himself in the community of scholars.

Things only got worse for Cavalieri and Torricelli's friends and students, those who, under normal conditions, would have carried on their trailblazing work. This generation included many talented mathematicians, but none of them (with a single exception) found it possible to continue down the road laid out by his teachers. Take, for example, Cavalieri's student at Bologna Urbano d'Aviso (b. 1618), who wrote an admiring biography of his teacher, but who, when it came to mathematics, contented himself with authoring an elementary textbook on astronomy. Another Cavalieri student, Pietro Mengoli (1626–84), succeeded Cavalieri to the mathematics chair at Bologna and was a subtle and talented mathematician. But he was also a conservative one, who avoided indivisibles and later retreated entirely from mathematics to engage in solitary religious meditations. Giannantonio Rocca (1607–56), Cavalieri's friend and the man who warned him against publishing his polemic against Guldin as a dialogue, was also noted for his skill as a mathematician, and Cavalieri even included some of his results in his *Exercitationes*. But Rocca himself never published a line on indivisibles.

The case is much the same with Torricelli's associates. Vincenzo Viviani (1622–1703) was, along with Torricelli, Galileo's friend and companion in his final years. The two worked together in Florence after Galileo's death, and Viviani ultimately succeeded Torricelli as mathematician to the Medici court. Viviani always considered himself Galileo's student and intellectual heir, and wrote a biography of the Florentine that serves as the basis for all modern ones. But when it came to mathematics, Viviani's work was almost entirely in the classical mold: he translated the ancient classics and issued new editions of Apollonius's *Concis* and Euclid's *Elements*, but only very rarely did he refer to indivisibles, and then only to repeat well-known results, such as the quadrature of the parabola. By his later years, he seemed to have given up even on that: when Leibniz in 1692 published a solution to certain mathematical problems left standing by Galileo, Viviani harshly criticized him for making use of infinitesimals. By this time, it seems, even Galileo's students in Italy had accepted, and perhaps internalized, the ban on the infinitely small.

Antonio Nardi, another of Torricelli's friends, wrote extensively on mathematics and in support of the method of indivisibles—but never published, despite his repeatedly stated intent to do so. All that remains of his work are thousands of pages in the archives of the Biblioteca Nazionale Centrale in Florence, of which not a word ever saw light. And then there was the case of Michelangelo Ricci, a student of Torricelli and Castelli in Rome in the 1630s who, remarkably, rose to become a cardinal of the Church. Ricci was a talented and well-regarded mathematician who was also, as his letters show, an admirer of Galileo, Cavalieri, and Torricelli and an enthusiastic practitioner of the method of indivisibles. But he, too, kept his mathematical preferences under wraps and published nothing on the subject.

The silence of Mengoli and Nardi, Viviani and Ricci, tells the story of the slow suffocation and ultimate death of a brilliant Italian mathematical tradition. Cavalieri, the oldest of Galileo's mathematical disciples, was fortunate enough to secure a university position in the 1620s, when the Galileans were ascendant in Rome. The younger Torricelli encountered a far harsher environment in the 1630s, but was saved from oblivion by the remarkable stroke of luck that summoned him to Arcetri in time to be appointed Galileo's successor. But for those who wished to follow in their footsteps, the enmity of the Society of Jesus ensured that no such miracles were at hand. No city or prince wished to risk the wrath of the Jesuits, and as a result, no university chairs or positions of honor at princely courts were in the offing for supporters of the infinitely small. So they kept their silence, corresponding among themselves and with mathematicians abroad, but never publishing their work or drawing attention to themselves. And once they, too, passed from the scene, there was no one left in Italy to carry the torch of the infinitely small.

THE LAST STAND OF THE INFINITELY SMALL

Before the champions of the infinitely small in Italy gave up the fight and conceded defeat to their enemies, they made one final stand in defense of their mathematical approach. It was conducted with wit and a

fearless spirit by the last Italian mathematician to advocate infinitesimals openly: Brother Stefano degli Angeli, of the order of the Jesuats of
St. Jerome. Angeli was born in Venice and entered the Jesuat order at a
young age. His intellectual gifts were apparently recognized early on,
because at the age of twenty-one he was sent to the Jesuat house in Fer-
rara to teach literature, philosophy, and theology. After a year or so, possibly due to ill health, he was moved again, this time to Bologna. There
he met the man who would help shape the rest of his life and career: the
prior of the Bologna house, his Jesuat brother Bonaventura Cavalieri.

When they met in the mid-1640s, Cavalieri was already a famous
man in mathematical circles, known as the father of the method of indivisibles. He was laboring on his response to Guldin's critique, but he was
also in bad health, suffering severely from the gout that would take his
life in 1647. In Angeli he found a friend and a disciple, one who embraced his mathematical approach with enthusiasm, and soon showed
himself a talented mathematician in his own right. It is easy to imagine
the two of them together, cloaked in the white habit and leather belt
of their order, the middle-aged man and his young disciple walking daily
through the busy streets of Bologna from the Jesuat house to the ancient
university. Were they speculating, as they walked, about the composition of the continuum or a new approach for calculating the area inside
a spiral? Were they debating the best response to Guldin and how to
present it? Or were they lamenting the latest salvo from the Jesuits? We
will never know, of course, but it is more than likely that all these topics
came up between them. What we do know is that the two formed a
strong bond, and that Angeli came to see himself as the guardian of Cavalieri's legacy. When in the last months of his life Cavalieri became too ill
to attend to the publication of the *Exercitationes*, it was Angeli who made
the final edits and saw the book through the press.

After Cavalieri's death Angeli was transferred again, likely at his own
request, and spent the next five years as rector of the Jesuat house in
Rome. It was an impressive promotion for a young man who was no more
than twenty-four, and it was no doubt aided by the strong support of his
late mentor. Angeli was, at the time, already an accomplished mathematician, so it is significant to note that during his entire stay in Rome he
did not publish a thing. We are already familiar with this pattern from
the experience of Torricelli, who spent the decade of the 1630s in Rome

deeply engaged in mathematics but who began publishing only when he was safely ensconced in the Medici court in Florence. The Eternal City, world headquarters of the Society of Jesus and home of the Collegio Romano, was not a place in which one could freely advocate the doctrine of the infinitely small.

But in 1652, Angeli was transferred back to his native city of Venice, where he was appointed provincial councilor (*definitore*) for his order. It would have been a welcome move, for Venice was a very good place for one seeking shelter from the long arm of the Society of Jesus. This is because back in 1606 the city got into a dispute with Pope Paul V over its right to try and punish ecclesiastics, and the Pope, enraged that the city leaders were infringing on his authority, excommunicated the city. The Venetian Senate, however, was undeterred: it demanded that the city's clergy continue to administer the sacraments despite the interdict, and the vast majority of the priests complied. The Jesuits, ever loyal to the Pope, did not, and were consequently expelled from the city. City and Pope reconciled the following year, but the Jesuits remained banned from Venice for the next fifty years. The Black Robes were finally allowed back in 1656, but even after their return, their influence in Venice remained limited. Angeli took full advantage: protected by the leaders of his own order and by a vigilant Venetian Senate that was still suspicious of the Jesuits, he was free to show his true colors, and began publishing on the method of indivisibles.

When Angeli entered the battle over the infinitely small, he did so with élan and a flair that had not been seen in decades. Cavalieri had tried to appease his critics by straying as little as possible from the classical canon, and later gave up on his provocative anti-Guldin dialogue. Torricelli simply refused to engage the critics of his method, and the others, from Nardi to Ricci, never published their views at all. But Angeli charged into the fray like an avenging angel, determined to strike back at the Jesuits for the stranglehold that was slowly suffocating the method he held dear. His first broadside was included in an "Appendix pro indivisibilibus," attached to his 1658 book *Problemata geometrica sexaginta* ("Sixty Geometrical Problems"), and it was aimed directly at Mario Bettini.

In defending indivisibles, Angeli ridiculed Bettini's discussion of a paradox presented in Galileo's *Discorsi*, in which the circumference of a bowl is shown to be equal to a single point. "Father Mario Bettini of the Society of Jesus," Angeli wrote, is "a man who because he was the author of the Apiary can be called The Bee." This is appropriate, he continues, because "just as a bee both makes honey and stings, so does Bettini: he makes honey, in teaching the sweetest doctrine, but he stings what is, according to him, wrong in mathematics." Unfortunately, Bettini is "an unlucky bee." Although he "uses his sting to fend off indivisibles, he is nevertheless in danger," because, as Angeli shows in detail, Galileo's paradox proves Bettini's position untenable.

Angeli's comparison of Bettini to a confused bee is mocking enough, but he is not yet done with the Jesuit. He quotes the passage in which Bettini calls the method of indivisibles "counterfeit philosophising" ("similitudinem philosophantium"), and exclaims, "far, far be it from me to wish to make my geometrical theorems useless." Seeing an opening, Angeli pounces: "Note here, reader, how this author, on falling in with indivisibles, cries out as if he were met with demons: Far, far be it from me, etc." Bettini is here a hysterical exorcist trying to fend off demonic indivisibles with furious incantations. But as to substance, Angeli concludes, "he adds nothing new but spite."

The hyperbolic Bettini was perhaps easy quarry, but neither was the more formidable Tacquet spared Angeli's sharp pen. In the preface to his *De infinitis parabolis* of 1659, Angeli describes how a few days after the publication of his previous book, in which he faced down the Jesuit Bettini, he wandered into the Venetian bookstore Minerva. There he came across *Cylindrica et annularia*, the work of another, "most deserving mathematician of the same Society." Flipping through the pages, he by chance came across a passage in which the author "carps on indivisibles," claiming that they are neither legitimate nor geometrical. Angeli claims that he had never previously heard of the book or known of its critique of indivisibles, but that is highly unlikely. He was extremely well versed in the mathematical output of his contemporaries, and later in the preface, he cites the Frenchmen Jean Beaugrand and Ismael Boulliau, the Englishman Richard White, and the Dutchman Frans van Schooten, as well as his fellow Italians. It stretches credulity that he was not familiar with the work of Tacquet, the leading Jesuit mathematician of the day, or Tacquet's views on indivisibles, until he stumbled upon them in a Venetian bookstore. Angeli's plea of ignorance is rather a rhetorical pose, aimed at presenting him as an impartial scholar reacting to outrageous claims made by Bettini and Tacquet. The long and bitter history that had pitted him against the Jesuits for decades is left unmentioned.

Angeli goes on to claim that there is nothing particularly troubling in Tacquet's criticism of indivisibles. His arguments are old, he writes, and were already raised by Guldin and satisfactorily answered by Cavalieri years ago. But Tacquet did provide Angeli with an opportunity to proclaim how influential the method of indivisibles had become by the late 1650s. "Who does this reasoning convince?" Tacquet asks rhetorically, pointing to what he considered the inherent implausibility of the method of indivisibles. "Whom does it convince?" Angeli repeats incredulously. Everyone, he responds, except the Jesuits.

Angeli here is trying to turn the tables on the Jesuits: rather than the indivisiblists being a lonely and diminishing band under attack from powerful enemies, it is the Jesuits who are lone holdouts against a method that is being universally accepted. Indeed, at first reading, the list cited by Angeli is impressive, and appears to support his case. But a closer inspection tells a very different story: Yes, Beaugrand, Boulliau, White, and van Schooten did indeed adopt Cavalieri's method, but they resided in faraway lands, north of the Alps. Of the three Italians whom Angeli cites—Torricelli, Rocca, and Raffaello Magiotti—only Torricelli had in fact published on indivisibles, whereas Rocca and Magiotti had remained mum; and in any case, by 1659 all three were dead. Despite his protestations to the contrary, Angeli was, in his own land, alone.

Satisfied with his rhetorical salvos, Angeli then confronts Tacquet's dark warning that unless it were destroyed first, the notion that the continuum is composed of indivisibles would destroy geometry. Cavalieri insisted that the question of the composition of the continuum was irrelevant to the method of indivisibles, and Angeli here follows his teacher—but only up to a point. Like Cavalieri, he, too, argues that Tacquet was wrong, and that "even if the continuum is not composed of indivisibles, the method of indivisibles nevertheless remains unshaken." But he adds a twist: "if in order to approve the method of indivisibles, the composition of the continuum from indivisibles is necessarily required, then certainly this doctrine is only strengthened in our eyes." In other words, unlike his cautious teacher, Angeli is perfectly willing to accept that the continuum is indeed composed of indivisibles. The power and effectiveness of the method of indivisibles is proof enough of its correctness, and if it leads to the conclusion that the continuum is composed of indivisibles, then that, too, must be correct. The fact that this doctrine leads to contradictions and paradoxes bothers him not at all.

Angeli, the flamboyant Jesuat, took on the Jesuits like no one had dared since the days of Galileo himself. He called them names, ridiculed their exorcism practices, and pretended never to have heard of the most illustrious mathematician among them. But nothing demonstrated the clash between Jesuit and Jesuat like their contradictory approaches to the question of the composition of the continuum. For the Jesuits, the notion that the continuum is composed of indivisibles led to paradoxes, and on that account alone it must be banned from mathematics. A method based on it, even if effective and fruitful, was unacceptable because it undermined the very reason for which mathematics was studied: for its pure logical structure. Angeli's view was the exact opposite: because the method of indivisibles was effective, he reasoned, its underlying assumptions must be true, and if they involved paradoxes, then we would just have to live with them. One approach emphasized the purity of mathematics, the other emphasized practical results; one approach insisted on absolute perfect order, the other was willing to coexist with ambiguities and uncertainties. And never the twain shall meet.

THE FALL OF THE JESUATS

Thanks to the protection of his order and the Jesuit-unfriendly Venetian Senate, it appeared that Angeli would get away with his open defiance of the Society of Jesus. He kept up his work, and over the next eight years he published an additional six mathematical books, in all of which he used and advocated the method of indivisibles. His greatest triumph came in 1662, when he was appointed to the chair of mathematics at the

University of Padua, a position once held by Galileo. The Jesuits, so powerful elsewhere in Italy, could only fume as the upstart Jesuat was raised to one of the most prestigious mathematical posts in all Europe. They never responded to his taunts, or denounced him openly, but quietly, patiently, bided their time.

The Jesuits were in a tight spot. As long as Angeli continued with his insolence, there was always a danger the forbidden doctrine might be revived in Italy and their decades-long campaign would come to naught. But what could they do? Angeli was safe in Venice, and if they ever thought they could persuade the authorities there to silence him, then surely his appointment to the Padua professorship showed them this was unlikely. So the Jesuits changed tactics: in Venice they might be of little account, but in Rome they were still ascendant. And so, in order to silence the last Italian voice advocating the infinitely small, they turned to the papal Curia.

The evidence for what happened next is circumstantial, the documents relating to the events buried to this day deep in the Vatican archives. But what we do know is this: On December 6, 1668, Pope Clement IX issued a brief suppressing three Italian religious orders: one was a community of Canons Regular, residing on the island of San Giorgio in Alga, in the Venetian lagoon; the second were the Hieronymites of Fiesole, a popular order that, at its height, had forty houses across Italy; the third were the Jesuats of St. Jerome. As the brief put it, "no advantage or utility to the Christian people was to be anticipated from their survival."

The Canons Regular was a tiny community, and confined to a single Venetian island, and it is plausible that the bureaucrats at the Vatican indeed came to the conclusion that it served no real purpose. The Hieronymites of Fiesole was a much larger order, but in their case the term *suppression* is misleading. For while it is true that they ceased to exist independently, the order was not in fact dissolved, but rather merged with its sister order of the Hieronymites of Pisa; the houses themselves continued to exist much as before. But for the Jesuats of St. Jerome the suppression was a death sentence: from one day to the next, the order simply ceased to exist, its houses dissolved and its brothers dispersed. It was a stunningly violent and unexpected end to an old and venerable order. Founded by the Blessed John Colombini in 1361 to tend for the poor and the sick, it had survived for exactly three centuries and seven years.

The official reason given, and the one quoted in all public sources today, is that "abuses had crept into the order." But this explanation is no more helpful than the claim that the order's survival served no purpose. Some scholars note that the Jesuats were often referred to as the "Aqua-vitae Brothers," a designation that may suggest lax morals and loose living. But this was far from the case: the nickname was given to them because of their dedication to treating victims of the plague, which they did by administering an alcoholic elixir produced in their houses. There is no evidence that the Church hierarchy objected to the Jesuats' medical practices or tried to put an end to them.

In fact, by all indications the Jesuats were a flourishing order. Pope Gregory XIII (1572–85), who patronized the Jesuits, also supported the Jesuats, and entered their founder, the Blessed John Colombini, in the official Church calendar, fixing July 31 as his feast day. The order expanded rapidly in the sixteenth and seventeenth centuries, and established dozens of houses across Italy. They must have been popular with the upper classes in Italian cities, since both the Cavalieris of Milan and the Angelis of Venice saw fit to send their gifted sons to be educated in the order. The fact that two members of the order occupied academic chairs in Bologna and Padua, among the most prestigious universities in Europe, added intellectual luster to the Jesuats that few orders could match. Although it is not easy to learn what life was like inside the Jesuat houses, nothing that we know suggests moral decrepitude. Cavalieri's letter to Galileo in 1620 about life in the Jesuat establishment in Milan, in which he complained about being besieged by old men who expected him to study theology, does not give one a sense that this was in any way a party house. Nor does one get that feeling about the house in Bologna, where Cavalieri resided for the last eighteen years of his life, ill with gout, and where he engaged in mathematical discussions with the young Angeli. The inescapable impression is that these were establishments with a serious focus on academic learning and religious ministry, and the rapid advancement of both Cavalieri and Angeli to positions of authority in the order indicate that intellectual achievement was highly prized. Before 1668 the Vatican generally saw no reason to intervene in

Jesuat affairs, except in 1606, when, for the first time, it allowed clergymen to enter the order—a change that suggests a rise rather than a di-
minishment in the order's standing. There is nothing in all this that
would explain why this old and venerable brotherhood was singled out
for sudden annihilation.

But the Jesuats did stand out in one way: they counted among their
members the most prominent Italian mathematicians promoting the
doctrine of infinitesimals. First Cavalieri and then Angeli, each in turn,
was the leading advocate for indivisibles in his generation, and both
received the full backing of their order. Not only were they promoted
rapidly through the ranks, but many of their books were personally approved by the general of the Jesuats. Inevitably, when Angeli and Cavalieri entered into a bitter conflict with the Jesuits over the infinitely
small, the fight became not just their own, but that of their order as
well. Whether by design or happenstance, the Jesuats of St. Jerome
became the chief obstacle for the Jesuits in their drive to eradicate the
infinitely small.

It is possible that if the Jesuits had found a way to silence Angeli
while leaving his brothers in peace, they would have done so. But it is
just as likely that they were eager to make an example of the smaller order, a warning to all those within the Church who would dare challenge
the Society of Jesus. In the end, the result was the same. Unable to persuade the Venetian authorities to discipline the insolent professor, they
turned instead to the papal Curia in Rome, where their influence was
decisive. They could not punish Angeli directly, so they let their fury
rain on the order that sheltered him and his late teacher. When faced
with the wrath of the mighty Society of Jesus, the Jesuats never stood a
chance. The order that had survived three hundred years of political and
religious upheaval, whose brothers administered the waters of life to
victims of the plague, and two of whose members rose to the heights of
mathematical distinction, simply evaporated at the stroke of a papal pen.

Surreally, the man at the eye of the storm remained unmoved—at
least geographically. Although the brotherhood that had been his home
since his youth had suddenly dissolved around him, Angeli was still professor of mathematics at the University of Padua, and still protected by
the Venetian senate. He remained in Padua for the next twenty-nine
years, until his death in 1697. But even though he still professed himself

an admirer of Galileo, and although he had previously published no fewer than nine books promoting and using the method of indivisibles, Angeli did not publish a single word on the topic ever again. The Jesuits had won.

TWO DREAMS OF MODERNITY

By the 1670s the war over the infinitely small had ended. With Angeli at long last silenced, and all the Jesuits' rivals driven underground or melting away, Italy was a land cleansed of infinitesimals, and the Jesuits reigned supreme. It was, for the Society, a great triumph, and it came at the end of a difficult campaign that had claimed many victims along the way. Some of them were famous in their day, such as Luca Valerio and Stefano degli Angeli, but many more will remain forever anonymous. The cold hand of the Society of Jesus drew a curtain over this lost generation of Italian mathematicians and left them in the dark.

The Jesuits did not fight this battle out of pettiness or spite, or merely to flex their muscle and humiliate their opponents. They fought it because they believed that their most cherished principles, and ultimately the fate of Christendom, were at stake. The Jesuits were forged in the crucible of the Reformation struggle, which saw the social and religious fabric of the Christian West tearing at the seams. Competing revelations, theologies, political ideologies, and class loyalties were all vying for the minds and souls of the people of western Europe, leading to chaos, hunger, pestilence, and decades of warfare. The one and only Truth of the ancient Church, which had united Christians and given purpose to their years, had suddenly disappeared amid the clamor of rival creeds. Reversing this catastrophe, and ensuring that it never recurred, was the overriding purpose of the Society of Jesus from the day of its founding by Ignatius of Loyola.

The Jesuits pursued this goal in many ways, but always with energy, skill, and determination. They became expert theologians dedicated to formulating a single religious truth, and expert philosophers to support their theology. And they founded the largest educational system the world had ever seen, in order to disseminate knowledge of these truths far and wide. They were the engine of the Catholic revival in the second half of the sixteenth century and played a key role in halting the spread of the Reformation and reversing some of its gains.

But the Jesuits were confronted with a pesky problem: different opinions were everywhere, and every religious or philosophical doctrine was seemingly in contention between different authorities. Except, that is, in mathematics. This at least was the opinion of Christopher Clavius, who began advocating for the field at the Collegio Romano in the 1560s and '70s. In mathematics, and especially in Euclidean geometry, there was never any doubt, Clavius argued, and he ultimately made mathematics a pillar of the Jesuit worldview.

It is because of their deep investment in mathematics, and their conviction that its truths guaranteed stability, that the Jesuits reacted with such fury to the rise of infinitesimal methods. For the mathematics of the infinitely small was everything that Euclidean geometry was not. Where geometry began with clear universal principles, the new methods began with a vague and unreliable intuition that objects were made of a multitude of minuscule parts. Most devastatingly, whereas the truths of geometry were incontestable, the results of the method of indivisibles were anything but. The method could lead to error as often as to truth, and it was riddled with contradictions. If the method was allowed to stand, the Jesuits believed, it would be a disaster for mathematics and its claim to be a fount of incontestable knowledge. The broader implications were even worse: If even mathematics was shown to be riddled with error, what hope was there for other, less rigorous disciplines? If truth was unattainable in mathematics, then quite possibly it wasn't attainable anywhere, and the world would once again be plunged into despair.

It was to avoid this catastrophic outcome that the Jesuits pursued their campaign against infinitesimals. But were the mathematicians in Italy who championed the method of indivisibles truly dangerous individuals who would revel in overthrowing authority? This hardly seems likely. Galileo and Cavalieri, Torricelli and Angeli were, after all, academics and professors, hardly a breed of men inclined to overthrow civilization. Galileo may have been a flamboyant individualist, but he was no opponent of order, as he made clear when he chose to leave republican Venice to take a post at the court of the Grand Duke of Tuscany. Cavalieri was a sedate cleric and professor who left the city of Bologna

only once in the last eighteen years of his life, and Torricelli, after settling in Florence, did his best to avoid conflict with his critics. Angeli
undoubtedly showed a great deal of spirit in taking his last stand for indivisibles, but it would be exceedingly hard to describe him as a subversive. He was, after all, a cleric and a professor who relied on the protection
of his ancient order and the Venetian Senate to keep his enemies at bay.
It would indeed be difficult to find anyone among the proponents of the
infinitely small who justified the Jesuits' ferocious reaction to the doctrine, or their fears of its implications.

So were the Jesuits simply wrong to fear the proponents of the infinitely small? Not exactly. For although it is true that the Galileans were
not social subversives, it is also true that they stood for a degree of freedom that was unacceptable to the Jesuits. Galileo was a brilliant public
advocate for the freedom to philosophize ("*libertas philosophandi*"), by
which he and his associates meant the right to pursue their investigations wherever they led. He openly mocked the Jesuits and their reverence
for authority, writing that "in the sciences the authority of thousands of
opinions is not worth as much as one tiny spark of reason in an individual man." Not only did Galileo argue that when Scripture and scientific
fact collide, it is the interpretation of Scripture that must be adjusted,
but he publicly transgressed the authority of professional theologians.
Not surprisingly, the Jesuits were furious. It was precisely the kind of
transgression they believed could lead to chaos.

Galileo was the chief public spokesman of his group, but his fellow
Linceans, his students, and his followers shared his views. All of them
believed in the principle of *libertas philosophandi* and saw in the trial
and condemnation of their leader a monstrous crime against the freedoms they cherished. For them, the Jesuit quest for a single, authorized,
and universally accepted truth crushed any possibility of philosophizing
freely. By championing the mathematics of the infinitely small, they
were taking a stand against the Jesuits' totalitarian demand that truth be
officially sanctioned.

The core conflict between the Jesuits and the Galileans was on the
questions of authority and certainty. The Jesuits insisted that truth
must be one, and in Euclidean geometry they believed they had found
the perfect demonstration of the power of such a system to mold the
world and prevent dissent. The Galileans also sought truth, but their

approach was the reverse of that of the Jesuits: instead of imposing a unified order upon the world, they attempted to study the world as given, and to find the order within. And whereas the Jesuits sought to eliminate mysteries and ambiguities in order to arrive at a crystal-clear, unified truth, the Galileans were willing to accept a certain level of ambiguity and even paradox, as long as it led to a deeper understanding of the question at hand. One approach insisted on a truth imposed from above through reason and authority; the other pragmatically accepted the possibility of ambiguity and even contradiction, and sought to derive knowledge from the ground up. One approach insisted that the infinitely small must be banned, because it introduced paradox and error into the perfect, rational structure of mathematics; the other was willing to live with the paradoxes of the infinitely small as long as they served a powerful and fruitful method and led to deeper mathematical understanding.

Taking place at the dawn of the modern age, the struggle over the infinitely small was a contest between opposite visions of what modernity would be. On the one side were the Jesuits, one of the first modern institutions the world had seen. With rational organization and unity of purpose, they were working to shape the early modern world in their image. Theirs was a totalitarian dream of seamless unity and purpose that left no room for doubt or debate, a vision that has appeared time and again in different guises throughout modern history. On the other side were their opponents, which in Italy were the friends and followers of Galileo. They believed that a new age of peace and harmony would be brought about not by the imposition of absolute truths, but through the slow, systematic, and imperfect accumulation of shared knowledge and shared truths. It was a vision that allowed for doubt and debate, freely acknowledging that some mysteries remained unsolved, but insisting that much could nevertheless be discovered through investigation. It opened the way for scientific progress, but also for political and religious pluralism and limited (as against totalitarian) government. This group, too, has had many different incarnations in the modern world, but their views are still recognizable in the ideals of liberal democracy.

In seventeenth-century Italy, the enemies of the infinitely small prevailed. The principles of hierarchy, authority, and the absolute unity of truth were affirmed, and the principles of freedom of investigation, pragmatism, and pluralism were defeated. The consequences for Italy were profound.

THE WELL-ORDERED LAND

For nearly two centuries, Italy had been home to perhaps the liveliest mathematical community in Europe. It was a tradition that stretched back to the counting houses of Italy's commercial hubs, and later encompassed professional and university mathematicians, forebears of today's academics. In the early sixteenth century, Cardano, Tartaglia, and their fellow "cossists" (as they were known) wagered money and possessions on their ability to solve cubic and quartic equations. Some decades later, classicists such as Federico Commandino and Guidobaldo del Monte idolized the ancients and produced new editions and translations of their work. And most recently, champions of the infinitely small (Galileo, Cavalieri, and Torricelli) pioneered new techniques that would transform the very foundations of mathematical inquiry and practice.

But when the Jesuits triumphed over the advocates of the infinitely small, this brilliant tradition died a quick death. With Angeli silenced, and Viviani and Ricci keeping their mathematical views to themselves, there was no mathematician left in Italy to carry the torch. The Jesuits, now in charge, insisted on adhering close to the methods of antiquity, so the leadership in mathematical innovation now shifted decisively, moving beyond the Alps, to Germany, France, England, and Switzerland. It was in those northern lands that Cavalieri's and Torricelli's "method of indivisibles" would be developed first into the "infinitesimal calculus" and then into the broad mathematical field known as "analysis." Italy, where it all began, became a mathematical backwater, a land in which there was no future for those seeking to pursue a mathematical career. In the 1760s, when the young mathematical prodigy Giuseppe Luigi Lagrangia of Turin sought to make a name for himself among the "great geometers" of the day, he was obliged to leave his homeland and travel first to Berlin and then to Paris. He succeeded, but his Italian roots were soon forgotten. To future generations he was and remains a Frenchman: Joseph-Louis Lagrange, one of the greatest mathematicians in human history.

The extinction of the Italian mathematical tradition was the most immediate result of the suppression of the infinitely small, but the Jesuit triumph had far deeper and more wide-ranging effects. Going back to the High Middle Ages, Italy had led all Europe in innovation—political, economic, artistic, and scientific. As early as the eleventh and twelfth centuries, it was home to the first flourishing cities to emerge from the Dark Ages. These cities played a vital role in reviving the long-dormant commercial economy, and were also sites of lively political experimentation in different forms of government, from autocratic to republican. In the thirteenth century, Italian merchants became Europe's first and wealthiest bankers, and beginning in the middle of the fourteenth century, Italy led the way in an artistic and cultural revival that transformed Europe. Humanists from Petrarch to Pico della Mirandola, painters from Giotto to Botticelli, sculptors from Donatello to Michelangelo, and architects from Brunelleschi to Bernini, made the Italian Renaissance a turning point in human history. In the sciences, Italians from Alberti to Leonardo to Galileo made crucial contributions to human knowledge and opened up new vistas of investigation. As a land of creativity and innovation, it is fair to say, Italy had no peer.

All this, however, came to an end around the close of the seventeenth century. The dynamic land of creativity and innovation became a land of stagnation and decay. The thriving commercial hubs of the Renaissance became marginal outposts in the European economy, unable to match the rapid expansion of their northern rivals. Religiously, the Italian peninsula came under the sway of a conservative Catholicism, in which no dissent from papal edicts was permitted and no other sect or belief was allowed a foothold. Politically, Italy was an amalgam of petty principalities ruled by kings, dukes, and archdukes, and by the Pope himself. With few exceptions, all were reactionary and oppressive, and all forcefully stifled any hint of political opposition. In the sciences, a few brilliant men, such as Spallanzani, Galvani, and Volta, worked at the forefront of their disciplines and were admired by colleagues throughout Europe. But these few exceptions only emphasized the overall impoverishment of Italian science, which in the eighteenth century was but an appendage to the flourishing Parisian science. By 1750 there was little trace of the bold spirit of innovation that had characterized Italian life for so long.

It would be an exaggeration to attribute all these developments to the defeat of the infinitely small in Italy in the late seventeenth century. There were many causes for Italian decline—political, economic, intellectual, and religious—but it is undeniable that the struggle over the infinitely small played an important role among them. It was a key site in which the path of Italian modernity was fought over and decided, and the victory of one side and defeat of the other helped shape Italy's trajectory for centuries to come.

It did not have to be so: the struggle was a close one, and if the Galileans had won and the Jesuits lost, it is easy to imagine a quite different way forward for Italy. The land of Galileo would likely have remained at the forefront of mathematics and science, and may well have led the way in the scientific triumphs of the eighteenth and nineteenth centuries. Italy might have been a center of Enlightenment philosophy and culture, and the ideals of freedom and democracy could have resonated from the piazzas of Florence, Milan, and Rome, rather than the places of Paris and the squares of London. It is easy to imagine Italy's petty dynasts giving way to more representative forms of government, and the great cities of Italy as thriving hubs of commerce and industry, fully the equal of their northern counterparts. But it was not to be: by the late seventeenth century the infinitely small had been suppressed. In Italy, the stage was set for centuries of backwardness and stagnation.

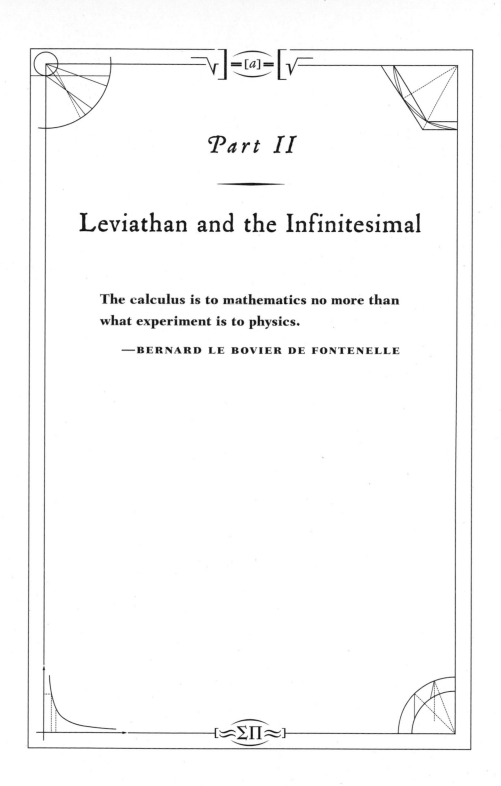

Part II

Leviathan and the Infinitesimal

The calculus is to mathematics no more than
what experiment is to physics.

—BERNARD LE BOVIER DE FONTENELLE

6

The Coming of Leviathan

DIGGERS

On Sunday, April 1, 1649, a group of poor men gathered with their families on St. Georges Hill, near the town of Kingston in Surrey, England. The hill was barren and seemed an unpromising locale for a new settlement. But the newcomers had come to stay: they had brought their belongings with them, and quickly set about building huts to shelter them from the elements. Then they began to dig. Day after day they continued digging, carving out trenches and planting vegetables on the rocky hill, while calling on others in the nearby towns to join them. "They invite all to come in and help them," noted one observer, "and promise them meat, drink, and clothes." They confidently predicted that "they will be four or five thousand within ten days," and while this proved overly optimistic, the community did attract newcomers, their numbers soon reaching several dozen families. And yet they went on digging.

As the community slowly grew, suspicion of the "Diggers" in the surrounding towns and villages grew along with it. "It is feared they have some design in hand," noted the same observer, and he was not mistaken. Digging trenches on a barren hill may seem like an innocent act to us, but things were different in seventeenth-century England. With their actions, the Diggers were asserting ownership and their right to cultivate enclosed lands that were owned or controlled by the local grandees. It was a calculated and open assault on the ownership rights of the propertied classes, and if their intentions were not sufficiently plain from their actions, the Diggers soon followed up with a pamphlet they distributed far and wide. "The work we are going about is this," they explained: "To dig up *Georges Hill* and the waste Ground thereabouts . . . that we may work in righteousness, and lay the Foundation of making the Earth a Common Treasury for All, both Rich and Poor . . . not one Lording over another, but all looking upon each other, as equals in the Creation."

Such a bold denial of the rights of private ownership would have been enough to send chills down a landowner's spine, then and now. But there was more: "that this Civil Propriety is the Curse, is manifest thus, Those that Buy and Sell Land, and are landlords, have got it either by Oppression, or Murther, or Theft." All private property was, according to this logic, stolen, and should by all rights be returned to its rightful owner: the people. True, the Diggers professed pacifism and made a point of disavowing the use of force to reclaim the land. But since several of their members were veterans of the English Civil War and its ravages, the "better sort" of people in Weyburn and surroundings were far from reassured. Having been labeled thieves and murderers, and their property rights denied, they were understandably alarmed. Fearing for their land and possessions, not to mention their lives and safety, they struck back.

As established members of society, they first turned to the authorities: Sir Thomas Fairfax, commander of the New Model Army, was stationed nearby, and the landowners appealed to him to remove the squatters. Fairfax was probably the most powerful man in England at the time, having led the Parliamentary forces to decisive victories over the Royalist armies of Charles I. A gentleman and a knight, Fairfax had little sympathy for the revolutionary demands of the Diggers, and the landowners expected him to take their side. Fairfax, however, equivocated: He arrived at St. Georges Hill with his troops and engaged in several discussions with the Diggers' leader, Gerrard Winstanley. Beyond this, however, he took no action. If the landowners had an issue with Winstanley's band, Fairfax informed them, they needed to take it up with the courts.

Though disappointed at Fairfax's response, the landowners did precisely that—and more. They charged the Diggers with sexual licentiousness, and prevailed on the courts to bar them from speaking in their own defense. Meanwhile, Francis Drake, lord of the nearby manor of Cobham, organized raids on the Diggers' settlement, ultimately succeeding in burning down one of their communal houses. Faced with a concerted legal and physical assault, the Diggers gave way. By August they had been forced to leave St. Georges Hill and move to a new location some miles away. When this new refuge also came under attack, they abandoned the land and largely dispersed. The landowners had won.

The drama at St. Georges Hill is one of the best documented attempts to subvert the established social order in early modern England. But it was not an isolated incident. Other digging colonies sprang up during the period, and other forms of protest, subversion, and even insurrection abounded. For, in the middle decades of the seventeenth century, from 1640 to 1660, England was a land in turmoil, and traditional institutions were in flux, if they had not disappeared altogether. Less than forty years after the death of the brilliant "Virgin Queen," Elizabeth I (1558–1603), her successor once removed, Charles I, was driven from London by Parliament, his armies defeated in the field and he himself imprisoned and ultimately executed. The Church of England, created by Elizabeth and her father, Henry VIII (1509–47), was effectively dissolved, its bishops forced into exile and its great cathedrals seized by rival Protestant churches. A Scottish army had invaded and, for a time, occupied the northern counties; while in Ireland, a Catholic uprising had laid waste to the lands of English lords and settlers, massacring many of them and forcing others to flee. Amid this national crisis, with the state decapitated, the official Church suppressed, the law of the land ignored, and print censorship removed, a multitude of groups emerged from the shadows, dedicated to turning the old world upside down. The Diggers of St. Georges Hill were but one of them.

THE LAND WITHOUT A KING

The causes of what is variously known as the English Revolution, the English Civil War, or simply the Interregnum, are debated by historians to this day. Political, religious, social, and economic causes are all cited, and indeed there is no doubt that all contributed in some way to the collapse of the English government in 1640. This much, however, is clear: ever since 1603, when James I (1603–25) of the House of Stuart succeeded Elizabeth I on the English throne, England's kings were increasingly at odds with Parliament, a body that represented large swaths of the English propertied classes. In part, this was a straightforward power struggle. Parliament, whose roots went back to the thirteenth century, had by the time of Elizabeth's reign acquired the exclusive right to levy taxes. Since raising and maintaining an army and a navy were by far the most costly undertakings of an early modern state, and could be financed only through taxes, this meant that the king could not pursue a foreign policy without Parliament's approval. Due to its control over state revenues, Parliament had the power to veto policies it did not like, and it did not hesitate to use it. As long as royal policies were acceptable to Parliament, there was little trouble. Such was the case with Elizabeth's long, expensive, and inconclusive war with Spain, which nevertheless retained broad popular support. But when James I made peace with Spain, and when Charles I decided to aid King Louis XIII of France in crushing the Protestant Huguenots, things changed. Parliament refused to authorize taxes to fund what it viewed as "godless" and "tyrannical" actions, making it impossible for the English king to effectively carry out his policies.

The Stuart monarchs found this situation intolerable. Only the king, they insisted, had the power to set policy and levy taxes, and Parliament's stranglehold over taxation was an illegal usurpation of royal power. They looked enviously at the French kings, who had humbled their own assembly, the Estates-General, and were successfully concentrating all power in their own hands. James I, perhaps the most scholarly of English kings, even penned a treatise entitled *The Trew Law of Free Monarchies*, in which he argued that kings ruled by divine right, and under no circumstances could the people legitimately resist royal decrees.

With Parliament growing ever more assertive and the Stuart kings ever more furious, a confrontation was inevitable: in 1629, Charles I dissolved Parliament and refused to call a new one. For the next eleven years he ruled alone, while the treasury was slowly drained and his freedom of action became increasingly restricted. Finally, in 1640, following a disastrous attempt to reform the Scottish church that led the two nations to the brink of war, Charles could hold out no longer and recalled Parliament. His intent was only to approve funds for the Scottish war and then quickly dissolve the unruly body. But Parliamentary leaders struck first: to prevent a repeat of Charles's "tyrannical" rule, they immediately passed a resolution that Parliament stay in session until it dissolved itself. It remained formally convened for the decade, and is known to history as the Long Parliament.

The constitutional crisis of 1640 was a clash of two fundamentally opposing views of the proper political order. The Stuart kings struggled mightily to establish an absolutist monarchy on the French model, in which all authority resided with the divinely sanctioned king. Parliament, meanwhile, stood for a constitutional monarchy (though the term had not yet been coined). Even the king, in its view, could not trample on the ancient rights of freeborn Englishmen. Royal power must be tempered and, when necessary, resisted by "the people," as represented in Parliament. Needless to say, parliamentary leaders never dreamed of including the lower classes and the poor among "the people" of England. Only property owners were represented in Parliament, so only they had the right to share power with the king. Even so, the Parliamentary party stood for a vast expansion of the political class in England, which was precisely what the Royalists were determined to prevent.

Today we are accustomed to thinking of constitutional issues, such as the proper balance of power between king and Parliament, as distinct from religious issues. But in seventeenth-century England, politics and religion were inseparable. Parliament's audacity in challenging the power of the king was derived in no small measure from the new Protestant faith, which taught that all men had equal access to God's grace through faith and prayer. Whereas in Catholicism, grace was channeled exclusively through ordained priests endowed with special powers, all Protestant denominations subscribed to the principle of the "priesthood of all believers." All men, accordingly, were "priests" before God, capable of receiving grace directly from Him. And if all men were equal before God, why should they accept the absolute rule of the king, who was, after all, a man like them?

This Protestant outlook, to be sure, did not mean that the Parliamentarians believed that "all men were created equal." Far from it. But it did mean that the divine right of kings—men elected by God to rule over the people—was harder to maintain in Protestant England than in Catholic lands, where royal supremacy was buttressed by the authority of the Church. Consequently, the English Parliament was far more aggressive in asserting its rights and powers than its continental equivalents. Whereas Parliament challenged the early Stuarts at every step, the French Estates-General and the Spanish Imperial Cortes soon wilted before the divinely sanctioned authority of their kings.

The intertwining of politics and religion meant that the constitutional struggle between king and Parliament in England was also a religious struggle over the proper forms of worship and their meaning. The Church of England was a compromise reached after sharp swings between radical Protestantism and conservative Catholicism. Under the Elizabethan settlement, the Church retained the Calvinist theology of the radicals but combined it with an institutional structure and liturgy hardly distinguishable from Catholicism. Alone among Protestant denominations, Anglicanism retained bishops, a strict church hierarchy with the king at its apex, and solemn rituals in grand cathedrals conducted by resplendently attired Church grandees. Anglicanism was an uncomfortable marriage of two very different notions of faith and community, but it allowed each of the competing factions to emphasize its own favored aspect of the compromise. Parliamentarians, broadly speaking, emphasized the Calvinist theology with its egalitarian implications; the kings, by contrast, favored the hierarchic, Catholic-like forms. As James I famously put it, "No Bishop, no King!"

By 1640 the rift between Parliament and the king had deepened to the point where the Anglican compromise no longer seemed tenable. The dominant factions in Parliament advocated the abolition of bishops and the entire Church hierarchy, in order to bring Anglicanism more in line with other Protestant denominations. The Stuart kings, meanwhile, flirted openly with Catholicism, and appeared inclined to abandon the Protestant experiment altogether and reunite with Rome. The religious conflict was inseparable from the political crisis, and made the latter ever more difficult to contain. With not just power but also faith and conscience hanging in the balance for both sides, the room for compromise between king and Parliament was shrinking fast. By 1640 it had all but vanished.

When the Long Parliament met in 1640, it began a systematic assault on the authority of both king and Church. It appointed an "Assembly of Divines" to come up with a plan for radical Church reforms, and it prosecuted, and ultimately executed, Charles I's chief minister, the Earl of Strafford. The dominant Parliamentary party, known as Presbyterians, advocated a Scottish-style Church government, meaning the abolition of bishops and their replacement with councils of lay elders ("presbyters"). Denying the king money to fund an army, they instigated a military crisis by inviting the Scots to invade the northern counties. By 1642, Charles had fled London and was raising an army in the north, with the aim of ousting the rebellious Parliament and reasserting his royal rights. Parliament countered by forming its own militia, and for the next two years civil war raged across England, with neither side gaining the upper hand. Large engagements were few and far between, but the sacking of manors and towns, disease, and devastation were plentiful, bringing misery to the British Isles.

In 1645, frustrated by the costly and inconclusive war, Parliament launched a radical military reform: the traditional local militias, commanded by the leading citizens of each town or shire, would be replaced by a true professional army commanded by professional soldiers appointed for their military prowess, not their social standing. Men would be recruited from all segments of society, and would be promoted for their skill regardless of their origin. Overall command was given to England's ablest soldiers, Sir Thomas Fairfax and Oliver Cromwell. The new creation was known as the New Model Army, and its impact was immediate and dramatic. In the Battle of Naseby, in June 1645, the Parliamentary forces routed the king's army and, soon after, captured Charles I himself.

Uncertain how to deal with their royal prisoner, the Parliamentarians vacillated. The Presbyterians wanted to reach an accommodation with him, retaining the monarchy but guaranteeing parliamentary powers and Church reforms. By this time, however, the Presbyterians were no longer the dominant party they had been five years earlier, but just one faction among many. Their power had been eclipsed by the more radical Independents, who denounced the Presbyterian Church hierarchy as no better than the Anglican or Catholic hierarchies, and insisted that each congregation rule itself. Even more radical were the Levellers, who advocated a "leveling" of the social order, and numerous sects, known as "enthusiasts," who claimed divine inspiration and predicted God's coming vengeance on the propertied classes. All these groups demanded that the captive king be held accountable for his oppression of the people, and many of their members advocated abolishing the monarchy altogether. Much to the Presbyterians' consternation, such views were especially prevalent in the New Model Army, the instrument of Parliament's victory.

Sensing division among his enemies, the king stalled for time. He played one group against the other and eventually managed to escape his captors and relaunch the war. All, however, to no avail: the New Model Army made quick work of the Royalist uprising, and by the end of 1648 the king was back in the army's custody. This time his enemies were determined not to let him slip through their grasp: When some in Parliament tried once again to negotiate with the king, the army purged the body of all but its most hardened radicals. This diminished "Rump Parliament," as it was known, then appointed fifty-nine commissioners who put the king on trial and quickly sentenced him to death. On January 30, 1649, Charles I was beheaded at the royal palace of Whitehall, in London, the only English king to have been tried and executed.

The beheading of Charles I did not end England's time of troubles, and it remained a land without a king for more than a decade thereafter. The contest between relative moderates in Parliament and radicals among the soldiery of the New Model Army continued, and control of the state alternated between the two factions. Finally, in 1653, as a way out of the stalemate, a new constitution was passed that declared Oliver Cromwell "Lord Protector of England," giving him absolute powers. He was likely the only man with sufficient authority and credibility to hold the state together, but even he found it a daunting challenge. Much like the French revolutionaries more than a century later, he chose to divert the passions of the internal crisis by taking England into a series of foreign wars—first against Scotland, then against the Dutch Republic, and eventually against Spain. Cromwell brought great energy and administrative skill to his new role. He maneuvered deftly between the demands of the radicals and moderates, and when political astuteness proved insufficient, he did not hesitate to use overwhelming force. Consequently, under the

Protectorate, England enjoyed a period of internal peace and stability, at least compared to what it had endured the previous decade.

But in September 1658, at the age of fifty-nine, Oliver Cromwell died. His son Richard, who succeeded him as lord protector, lacked his father's authority and the loyalty of the army, and was soon marginalized and forced to resign. With government controls once again removed, the same revolutionary groups that had so frightened the propertied classes in earlier years quickly reemerged. Winstanley's Diggers may have been gone for good, but many other groups, and unnumbered individuals, emerged to take their place. They varied enormously, from recognizable political parties such as the Levellers of London, to lonely prophets traveling the countryside in search of a following, to everything in between. All, however, rejected the rigid class system of their day and believed that God was omnipresent and accessible to anyone.

Some groups named themselves for their social agenda. The Levellers were the largest and most politically powerful group, pushing to follow up the removal of the king with egalitarian social reforms. The more moderate Levellers wished only to remove the social barriers between the propertied classes, whereas more radical factions sought a complete overthrow of the social order. Other groups were known for their "enthusiastic" religious stance, such as the Seekers and the Ranters, who denied the sinfulness of man and claimed that organized religion was a hoax designed to oppress the poor. The early Quakers, far from the dignified pacifists they would become in later years, were viewed as dangerous subversives, and Fifth Monarchy Men predicted the imminent end of the world, in which earthly hierarchies would be dissolved and God's elect reign.

To the established classes of England, whether nobles, gentry, merchants, or prosperous yeomen, it seemed as if the gates of hell had opened up and they were staring into the abyss. England, they believed, was on the verge of plunging right back into the darkest days of civil war. If central authority were not restored, the manors and estates of the countryside, the counting houses of London merchants, and the homes and property of gentlemen everywhere could all be swept away by the irresistible tide of an angry, religiously inspired mob.

Faced with a common threat of social revolution, the divided English elite set aside their bitter feud and joined forces. Even Presbyterians, who had fought for decades against royal "tyranny," now concluded that a king was better than anarchy. They sent feelers out to Charles II, son of the martyred king, who at the time was residing in Belgium with his court in exile. To their inquiry about his conditions for a restoration, Charles responded reassuringly: as king, he would work with Parliament, not against it, and he would not seek revenge against his erstwhile enemies. Even with these assurances, it still took the decisive intervention of the New Model Army to force the issue. In early 1660, General George Monck, commander of the army in Scotland, advanced on London, occupied the city, dissolved the Rump, and convened a new, moderate Parliament in its place. The new body immediately invited the king to return, and on May 25, 1660, Charles II landed in Dover. England was once more a monarchy.

The restoration of Charles II came as a great relief to the English upper classes. With a king once more on the throne, a legitimate government in power, and the Church of England reestablished, the threat of civil war and revolution receded and a degree of public order was restored. But the ghosts of the Interregnum continued to haunt the hearts and minds of Englishmen, and it soon became clear that not much, in fact, had been settled. The return to England of Charles II did not spell a victory for the absolutist dreams of the executed Charles I. In fact, it was not even a return to the status quo of the early 1600s. Before 1640, royal rule was taken for granted. Kings and queens might be overthrown and replaced by others, and the precise limits of their power might be contested, but there was no substitute for a divinely ordained monarch. Few people, if anyone, could imagine England ruled in any other manner.

The restored monarchy, however, was very different: it was not the inevitable continuation of royal rule from time immemorial, but the result of a careful political calculation by certain factions in Parliament and the army. Whatever his personal inclinations, Charles II well understood that his rule depended on keeping key parliamentary blocks, and the interests they represented, on his side. Because of this, he was a lesser creature than his royal forebears, shorn of much of the magical aura of kingship, and surviving as much by political acumen as by his claims to a mystical divine right. What shape the monarchy would take, and what place it would occupy in the life of the nation, were questions that would dominate political life in England for the next half century.

The debate over the character of the new regime was given extra urgency by the specter of the Interregnum, when Parliamentarian "Roundheads" had turned on Royalist "Cavaliers," Presbyterians on Anglicans, Independents on Presbyterians, Levellers on Independents, and Diggers on Levellers. To Englishmen of standing and means, it was a nightmare that must never be repeated. Even a Presbyterian minister deprived of his living by the returning king conceded that things were better under the king than when "we lay at the mercy and impulse of a giddy hot-headed, bloody multitude" whose "malice and rage was so desperate and giddy and lawless." The choice, as the diarist Samuel Pepys explained on the eve of the Restoration, was between the "fanatics" and "the gentry and citizens throughout England," and it was a battle that the "gentry and citizens" could not afford to lose. Even as they debated the shape and structure of the new Restoration regime, they were united by one paramount principle, which any government would be required to uphold: that the dark days of the Interregnum never return.

The state of affairs was in some ways reminiscent of the one the Jesuits found themselves in during the early decades of the Reformation. Then, it had been the ancient Church and the community of Western Christendom that were being rent apart by a multitude of heretical sects, each claiming sole ownership of divine truth. Just as in England a century later, the danger of social revolution was ever present, and memories of the peasant uprising in Germany and the Anabaptist republic in Münster would terrify the ruling classes of Europe for centuries to come. The parallels between their situation and the early Reformation were not lost on the Englishmen who lived through the Interregnum. Even Rev. Henry Newcombe, who had no love for Catholics, acknowledged the similarities, saying that during the Interregnum, England had been terrorized by a "Münsterian anarchy."

The Jesuits responded to the crisis of the Reformation by reaffirming the power of the Pope and the Church hierarchy as the only fount of absolute divine truth, and the foundation of an eternal universal order. In England, too, there were those who sought to reaffirm the absolute power of the sovereign and the state as the only means of maintaining order and keeping the fanatics at bay. Mostly they were royalists, courtiers and noblemen who had stood by Charles I and Charles II in exile and civil war and who believed that only the strong hand of a king could save the land.

One among them stood out. He was not an illustrious nobleman but an elderly white-haired commoner whose highest position in court was as mathematics tutor to the future Charles II. His appearance and position were unimposing, but his mind was reputed to be among the sharpest in Europe, and his philosophical writings were as bold and irreverent as they come. He fearlessly castigated clerics of all denominations as impostors and usurpers, and denounced the Pope and the entire Catholic hierarchy as the "Kingdom of Darkness." He despised the Jesuits, but nevertheless had this in common with them: he, too, feared social disintegration, and was convinced that the only answer was a strong central authority. His name was Thomas Hobbes, remembered today as the brilliant and provocative author of *Leviathan* and one of the greatest political philosophers of all time. He is less well remembered for another interest he shared with the Jesuits, one that he, like them, considered essential to his philosophy: mathematics.

THE BEAR IN WINTER

By the time he published *Leviathan*, his most famous and celebrated work, Thomas Hobbes was sixty-three years old, an old man by the standards of the day. Indeed, in some ways he must have seemed to his admirers and his enemies alike a man from another age. He was born in 1588 in the village of Westport, near Malmesbury, in Wiltshire, his premature delivery prompted by the shock his mother received when she got news of the Spanish Armada. As Hobbes told it in an autobiography he wrote in Latin verse toward the end of his life, she "did bring forth Twins at once, both Me, and Fear." That Hobbes was a fearful man we know from his own testimony, for he claimed to be afraid of the dark, of thieves, of death, and (with some justification) of persecution by his enemies. This might seem surprising coming from a man of Hobbes's maverick reputation, and there was certainly nothing timid about the way he presented his radical new philosophy, or his fearless attacks on the conventions and beliefs his contemporaries held dear. But in a deeper sense, Hobbes may have understood himself best, as his philosophy was indeed founded on fear: fear of disorder and chaos, of the "war of all against all," of a "Münsterian anarchy" that must be held at bay at all costs.

Hobbes's father, also named Thomas, was a country vicar apparently known more for his drinking and skill at cards than for his learning. "He was one of the clergie of Queen Elizabeth's time—a little learning went a great way with him and many other Sir Johns in those days," wrote Hobbes's friend and biographer John Aubrey, referring to a time when many of the country clergy were semiliterate, if that. In contrast, the younger Hobbes claimed that he himself was drunk only a hundred times in his entire life, or hardly more than once a year. When he did drink, he had "the benefit of vomiting, which he did easily; by which benefit neither his wit was disturbt longer then he was spuing nor his stomach oppressed." Aubrey, who reported this and calculated the frequency of Hobbes's intoxications, considered it a laudable record of sobriety, which it likely was in that age of stupendous drinking.

When Hobbes Senior was forced to leave Malmesbury after an altercation with a parson, his wealthier brother, Francis, took charge of young Thomas's education. For the next few years, Hobbes was tutored in Latin, Greek, and rhetoric, and at age fourteen he enrolled in Magdalen College, Oxford. He emerged six years later with a bachelor's degree in hand and an abiding distaste for Scholastic philosophy, which formed the core of the university curriculum at the time. He wanted to study the new astronomy and geography, he wrote years later, not the Aristotelian corpus, and to "prove things after my owne taste," not according to the narrow Aristotelian categories. His scorn for Aristotle, and his determination to go his own way, would last him a lifetime.

Shortly after graduating, Hobbes was hired as tutor and companion to the son of William Cavendish, soon to be named the Earl of Devonshire. It was a lucrative position for a bright young commoner with university training, and Hobbes likely took the post without a second thought. But the association with the Cavendish family would continue throughout his life, and did more to shape the course of his life and studies than

Thomas Hobbes, 1588–1679. This portrait by John Michael Wright dates from 1669 or 1670, when Hobbes was eighty-two years old.
(bpk, Berlin / Art Resource, NY)

anything he learned at Oxford. The Cavendishes were one of the great noble clans of England, and could trace their ancestry to the reign of Henry I, son of William the Conqueror. More recently, in addition to the traditional military and political services that great families were expected to provide the king, they had also distinguished themselves through their keen interest in the "new philosophy," as science was then known. Charles Cavendish (1594–1654), for instance, was a respected mathematician; his brother William (1593–1676), the Duke of Newcastle, maintained a laboratory on the grounds of his estate; and William's wife, Margaret (1623–73), was an acclaimed poet and essayist with a strong affinity for the natural sciences. Those Cavendishes who were not scholars or writers were patrons of the arts and sciences, and their country houses were centers of cultural and intellectual life. As a member of the Cavendish household, Hobbes gained access to the highest literary and artistic circles in the land. At Chatsworth and Welbeck Abbey, the Cavendish estates, he found the intellectual challenge and stimulation that he never experienced during his years at Oxford.

In joining the Cavendish household, Hobbes was following a well-trodden path for Renaissance intellectuals, for nothing provided the income, resources, or freedom to pursue one's interests like the patronage of a noble family. The great Italian artists and humanists (Leonardo, Michelangelo, Pico della Mirandola, to name but a few) had enjoyed the patronage of the Medicis of Florence, the Sforzas of Milan, and a long list of Renaissance popes. Even Galileo, who was already a famous man at the time, chose the life of a Medici courtier over a secure but mundane existence as university professor in Padua. In England, the polymath Thomas Harriot (1560–1621) had been a member of the households of Sir Walter Raleigh and then of Henry Percy, Earl of Northumberland; and Hobbes's contemporary the mathematician William Oughtred (1575–1660) was a tutor to the son of the Earl of Arundel.

But even as Hobbes chose the traditional path of patronage, other aspiring men of letters were seeking alternative routes to economic security. William Shakespeare (1564–1616) made an excellent living performing and selling his plays on the open market, and most playwrights of his day did the same—though rarely as successfully. The mathematician Henry Briggs (1561–1630) found a home in the newly established Gresham College of London, where he became the first professor of geometry in England and gave public lectures for a fee. Even Oxford and Cambridge, the notoriously conservative universities whose chief aim was to prepare young men for the cloth with a rigid medieval curriculum, occasionally opened their doors to more modern scholars. Briggs, for one, ended his days as the first Savilian Professor of Geometry at Oxford. Hobbes's choice of attaching himself to a noble household was not unusual for his day, but by the mid-1600s, when he published his most important works, it had come to seem rather old-fashioned. That, along with his advanced age and the fact that he had grown up in the glory days of Elizabeth's reign, set him apart from most of his friends and rivals.

Old-fashioned or not, the patronage of a noble clan still held many attractions, and Hobbes enjoyed them to the fullest. Three times between 1610 and 1630, Hobbes embarked on grand tours of the European Continent with his charges, young noblemen of the Cavendish family and its circle. He put these journeys to good use. While traveling in Italy in 1630, he called on Galileo, whom he admired, and whom he praised ever after as "the one who has opened to us the gate of natural philosophy universal, which is the knowledge of *motion*." In Paris he made the acquaintance of Marin Mersenne, the friar who was the nexus of the European "Republic of Letters," corresponding with scholars and communicating their queries, comments, and results to one another. Through Mersenne, Hobbes came into contact with the philosopher René Descartes, the mathematicians Pierre de Fermat and Bonaventura Cavalieri, and many others, becoming in effect a full-fledged member of the European intellectual world.

Hobbes made one other illustrious acquaintance through his connection with the Cavendish clan: for several years in the 1620s he served as personal secretary to Francis Bacon (1561–1626), the philosopher and great promoter of experimental science. Unlike some of his continental contemporaries, and Descartes in particular, Bacon believed that knowledge should be acquired through induction (the systematic accumulation of observations and experiments), not through pure abstract reasoning. Bacon was one of the leading jurists in England and had served as Lord Chancellor to James I until he was accused of corruption and impeached in 1621. In retirement he turned philosopher, and spent his days writing down his thoughts on natural science and its proper method. In fact, nearly all the works for which Bacon is remembered today date from the brief period in which Hobbes knew him, the years between his forced retirement and his death in 1626. Aubrey recounts how Hobbes would accompany Bacon on his walks around his estate, Gorhambury House, and write down the old man's thoughts. Bacon allegedly preferred Hobbes to all his other secretaries because Hobbes alone understood what he was transcribing. Hobbes's association with Bacon demonstrates the reach of the aristocratic connections made available to him through the Cavendish family, but it is not without irony: in later years, those who saw themselves as Bacon's true heirs and put his ideas into practice viewed his former associate Hobbes as their most dangerous enemy.

Another advantage that Hobbes enjoyed as a member of an aristocratic household was that he was under no pressure to publish anything. Shakespeare had to produce a steady stream of plays to make his living; Henry Briggs had to give public lectures; even Clavius at the Collegio Romano was expected to teach and author textbooks. But those who enjoyed the patronage of great families were rewarded mostly for being good company to their patrons, not for their productivity. This made for a comfortable life for a scholar, who could devote his time to contemplation and research, but it could also have strange consequences: Thomas Harriot, for example, was reputed to be one of the leading mathematicians in Europe, and modern-day studies of his manuscripts make clear that the reputation was well deserved. But he was a lifelong member of the Raleigh and Percy households, and as a result he never published a single page of the literally thousands of mathematical papers he left behind.

Under ordinary circumstances, this would very likely have been Hobbes's fate: over his decades as a Cavendish retainer, despite his widely acknowledged brilliance and his connections with leading intellectuals both in England and on the Continent, he published nothing, the only exception being a translation of the ancient Greek historian Thucydides's *Peloponnesian War*. It was Hobbes's sole publication well into middle age, and it seemed highly likely that nothing more would be forthcoming. He would have remained an obscure and shadowy figure, known today only to the most dedicated antiquarians. But in 1640, when he was fifty-two years old, Hobbes's comfortable world fell apart, and suddenly he started writing and publishing at a frantic pace. He did not stop until his dying day.

The crisis of 1640 struck the Cavendish household like a thunderbolt. Like most of the great noble clans of England, the Cavendishes were dedicated Royalists who remained unswervingly loyal to the House of Stuart throughout the Interregnum. The Parliamentary revolt was, to them, a simple commoners' rebellion that must be crushed by force, and they were quick to take up arms in defense of their king. William

Cavendish, the future Duke of Newcastle, and Charles Cavendish, son of the Duke of Devonshire, both held high command in Charles I's army in the early years of the civil war, and fared as poorly as the king's fortunes. Charles was killed in battle in 1643, and William was forced to flee to the Continent following the Royalist defeat at Marston Moor in 1644. He eventually made his way to Paris, where he joined other members of his household in Charles I's court in exile. Among them was Thomas Hobbes.

For Hobbes, the choice of the Royalist side in the civil war was a natural one. Although himself a commoner, he was an esteemed member of a noble household and had come to share the Cavendishes' social and political outlook. In 1640, at the first signs of trouble, he picked up and moved to Paris, where he joined a growing community of Royalist exiles. Comfortably settled, he quickly renewed his connections with Mersenne and his French correspondents. As the leading intellectual at the Stuart court, he was ultimately offered the post of tutor to the Prince of Wales, the future Charles II, but here, for the first time, he encountered the kind of opposition that would dog him for the rest of his life. Several leading courtiers objected to his appointment on the grounds that he was a materialist and an atheist, someone who would infect the future king with his heretical views. It was ultimately decided that Hobbes could become royal tutor, as long as he promised not to stray into philosophy or politics, and to stick only to the field of his expertise. That, of course, was mathematics.

If the loyal and steadfast Hobbes had come to inspire fear, and even revulsion, in the Stuart court, the reason was clear. By 1645, when the possibility of his appointment as royal tutor came up, he was no longer known simply as the humble house intellectual of the Cavendish clan, but rather as an unconventional and provocative philosopher whose views were likely to offend churchmen of all stripes as well as many dedicated Royalists. For, in 1642, shortly after arriving in France, Hobbes published his first political work, a learned tome called *Elementorum philosophiae sectio tertia de cive* ("The Third Part of the Elements of Philosophy, on the Citizen"). Written entirely in Latin, the book was aimed at professional philosophers, not royal courtiers, but enough about its contents filtered through to reach Charles I's advisers to put Hobbes under suspicion.

Most men in Hobbes's situation might have tried to reassure their critics, or at least refrain from giving further offense. The courtiers were, after all, his social betters as well as his allies in the struggle to restore the monarchy. But, as his critics soon discovered, Hobbes was not one to downplay his views or shy away from a fight. In 1647 he republished *De cive* (as the work was popularly known), and three years later he issued an English translation (*On the Citizen*), so that it could be better understood by his countrymen, both in England and at the court in exile. That same year, 1650, he issued two more English-language tracts, *Human Nature* and *De Corpore Politico, or the Elements of the Law*, which together explained his views on human nature and the political order that human nature made necessary. Finally, in 1651, he capped this torrent of creativity with his masterpiece, the work that made him one of the immortals of philosophy: *Leviathan*. By that time, as a consequence of this literary outpouring, Hobbes was persona non grata in the Stuart court. In 1652, having nowhere else to go, he left Paris and moved back across the Channel. And although he lived for another twenty-eight years, he never set foot outside England again.

"NASTY, BRUTISH, AND SHORT"

Leviathan is the child of the English Civil War, and in more ways than one. During the long decades of silence spent in the Cavendish household, Hobbes was quietly building up an elaborate philosophical system. It was supposed to have three parts, beginning with "On Matter" (*De corpore*), continuing with "On Man" (*De homine*), and concluding with "On the Citizen" (*De cive*). Given his track record, we may legitimately doubt whether under normal circumstances any of these treatises would have seen light, but the crisis of 1640 interrupted Hobbes's leisurely preparations. Instead of proceeding systematically through his philosophy, he now felt that it was the third part, on political life, that mattered most. With a newfound urgency, he finished up *De cive* and rushed it to print (though it is formally called "the third part"). He then quickly followed it up with the other political tracts culminating with *Leviathan*, which summarizes his overall views but focuses on politics. At a time when England was being torn apart by civil war, sedate discussions on the nature of matter had to make way for a prescription for creating a peaceful and stable state.

It was not just *Leviathan*'s timing, however, that marked it as the product of the civil war, but also, more deeply, the darkness of Hobbes's vision and the desperate solution he proposes. Behind every line in *Leviathan* lurks the specter of the social anarchy and fratricidal war that was convulsing England. Behind every crisp and stately phrase, behind every elegant philosophical argument, loom the unruly mobs bullying their betters; the great houses turned to ashes; the bloody battlefields of Marston Moor and Naseby; the murdered king. As Hobbes saw it, Parliament had removed the king and unleashed every form of political and social subversion. Order was replaced by chaos, and civic peace by a cycle of civil war that seemed to feed on itself to the point where it no longer seemed to matter who was fighting whom, or why. It was no longer the war of Parliament against the king, or of Presbyterians against Anglicans, but simply of all against all. The only way to end it, Hobbes believed, was to reinstate the sovereign, return the demons of anarchy and subversion to the pits of hell from which they had come, and seal them there forever. *Leviathan* would show how.

The first step toward ending the chaos of civil war was to understand what had led to it, and according to Hobbes, it was not the political and religious controversies that were raging in England, but something more fundamental: human nature. Men, Hobbes explains in *Leviathan*, are not particularly aggressive beings. All they desire is food, sex, some creature comforts, and a modicum of security to enjoy all these. The problem is that, without an established political order—what Hobbes calls a "Commonwealth"—men could have no security. What one person had worked for and acquired, another could come and take from him, and suffer no consequence. All men were therefore fearful of all their neighbors, and the only way they could achieve any measure of security was by gaining power over them. It is, in other words, fear that leads men to make war upon their neighbors. Unfortunately, Hobbes warns, power never leads to security, because once men acquire it, they inevitably seek more: "And the cause of this," he writes, "is not always that a man . . . cannot be content with a moderate power but because he cannot assure the power and means to live well, which he hath at present, without the acquisition of more."

If men are left to their own devices, according to Hobbes, fear of one's neighbors leads to war, war leads to more fear, which in turn leads to more war. Under such conditions there is no point in investing in the future, and life is a misery: "there is no place for industry, because the fruit thereof is uncertain, and consequently no culture of the earth, no navigation, nor building, nor instrument of moving and removing such things as require much force, no knowledge of the face of the earth, no account of time, no arts, no letters, no society, and which is worst of all, continual fear and danger of violent death." This is life in Hobbes's "state of nature," and he sums it up in what might be the most famous line in all political philosophy: in the absence of political order, human life is "solitary, poor, nasty, brutish, and short."

This was also, according to Hobbes, precisely the condition of England during the civil war. With the removal of the king, Englishmen had reverted to the state of nature, and were engaged in a "war of every man against every man." Men might give any number of reasons for waging war upon their neighbors: Presbyterians and Independents might accuse one another of erroneous religious doctrines; Levellers and Diggers might denounce the rich and claim that all men are created equal; Fifth Monarchy Men might proclaim that they were preparing the way for the Day of Judgment. But all these fancy claims were to Hobbes mere window dressing, because the real reason Englishmen were fighting one another was much simpler: fear. With the sovereign gone, people were left defenseless against the depredations of their neighbors, and to gain some measure of security, they resorted to attacking them first. The result was an endless cycle of violence. Hobbes had already seen it happen once, when he visited France with his Cavendish charge in 1610, shortly after the assassination of King Henri IV. The fear and disorder of those days left a deep impression on young Hobbes, who would never forget what happened to a land that deposed its sovereign. The Frenchmen of 1610 gave different reasons for their actions than the Englishmen of 1640, but that hardly mattered: in a land without a sovereign, Hobbes learned, every man lives in fear and makes war upon his neighbor.

But if eternal civil war is the natural state of human society, how can it be quelled? How can people gain the security for themselves and their families that is required if agriculture, commerce, the sciences, and the arts are to flourish? The answer, according to Hobbes, lies in a uniquely human attribute: reason. Animals are forever trapped in the state of nature, and some humans, too, such as "the savage people in many places of America." But reason gives men a choice. They can remain in the miserable state of nature, or they can recognize their unhappy condition, and rationally seek a solution that will lift them out of it. But once they choose to do so, according to Hobbes, all choices end, because reason will lead them to the only solution to their quandary: the Leviathan.

What is the Leviathan? It is much more than an absolute ruler, or even an absolute state. It is the literal embodiment of all the members of the commonwealth in one man: the sovereign. In their desperation to escape the state of nature, men conclude that the only way out is for each of them to give up his own free will and invest it in the sovereign. The sovereign consequently absorbs the individual wills of all the members of the commonwealth, and his actions are therefore their actions. This is the key. Men do not simply submit to the will of an overlord, subjugating their own will to his. Rather, whatever the sovereign wills is also the will of every one of his subjects. Every person in the commonwealth, Hobbes argues, will "own and acknowledge himself to be the author" of whatever the sovereign chooses to do. Under the Leviathan there can be no civil war, because the Leviathan embodies the wills of his subjects, and no one would will civil war. The end result is a perfectly unified body politic: "[It] is more than consent, or concord: it is a real unity of them all, in one and the same person."

"The multitude so united is called a COMMONWEALTH," Hobbes writes:

> This is the generation of that great LEVIATHAN, or rather (to speak more reverently) of that *Mortal God* to which we owe, under the *Immortal God*, our peace and defence. For by this authority, given him by every particular man in the commonwealth, he hath the use of so much power . . . that by terror thereof he is enabled to conform the wills of them all to peace . . .

And that, to Hobbes, is the very essence of the Leviathan: "one person, of whose acts a great multitude . . . have made themselves every one the author" in order that peace will prevail.

Hobbes's theory of the state is breathtaking in its audacity. He has no interest in discussing the different forces operating in human society or evaluating the different forms of political organization. Instead, with no qualifications or equivocation, he plunges ahead in a take-no-prisoners philosophical style. The problem of human society, he claims, is clear, and it is the perpetual war that exists in the state of nature. The solution is just as clear: the creation of the absolutist "Leviathan" state. Hobbes drives through his argument by sheer intellectual force, moving step by logical step, and leaving no room for dissent or contradiction: human nature leads to the state of nature, which leads to civil war, which leads to surrender of personal will, which leads to the Leviathan. Consequently, the Leviathan is the only viable political order. *QED*.

From the very beginning, many found Hobbes's Leviathan state abhorrent. Where would Parliament fit in? Where would the Anglican Church—or any church, for that matter? But even those critics who were repelled by his conclusions were hard pressed to find flaws in his arguments. For where exactly was Hobbes's error? His assumptions were sound, and each step seemed reasonable in itself: Yes, humans are acquisitive and self-interested. Yes, they compete with and fear one another. Yes, they are prone to attacking each other out of fear, and one attack leads inevitably to more. It all seems oh so reasonable, and few would be inclined to argue with each particular step. By the time a reader realizes where all this is leading, it is too late. Somehow, without ever taking a false or even dangerous-looking step, the reader unwittingly concedes that the only viable state is Hobbes's "living God," the Leviathan.

To many of Hobbes's contemporaries, this was a completely repellent conclusion, but such is the power of *Leviathan*'s reasoning that it proved extraordinarily difficult to point out where exactly it goes astray. Hobbes followed his deductions to their logical conclusions, whatever those might be, and carried his readers along for the ride. It was as if he were conducting a geometrical demonstration.

The Leviathan, composed of innumerable individuals united in a single will, is, to be sure, a beautiful thing. But bold as it is, and beautiful as it is, the Leviathan as a political organization is bound to give one pause. It is not just a powerful and centralized state as existed in Hobbes's time in France, where political opposition was difficult and state measures repressive. It is, rather, a state in which political opposition is literally impossible. Opposition to the sovereign by his subjects means that they willfully oppose their own will—a paradox and a logical impossibility. Indeed, in the Leviathan the subjects do not have the same relationship to the state as we understand it, because the Leviathan is not a political organization but a unified organic whole. It is a living being composed of the bodies of all its subjects, and a will entrusted to the sovereign alone. Hobbes says as much when he explains in the introduction that "the great LEVIATHAN, called a COMMONWEALTH . . . is but an artificial man, though of greater stature and strength than the natural." In a human body, a hand, or a foot, or a follicle of hair cannot oppose a human's will. In just the same way, the members of the commonwealth are simply components of the body of the commonwealth, and are incapable of opposing its will.

Nothing captures the true essence of the Hobbesian state better than the image that adorned the frontispiece of the early editions of *Leviathan* (and many later ones as well). Engraved by the French artist Abraham Bosse, it shows a peaceful land of hills and valleys, fields and villages, with a prosperous and orderly town in the foreground, where small and neatly arranged houses are dwarfed by a great Church. The eye, however, does not dwell on this peaceful scene, but is drawn to the figure that looms behind it: a giant king who towers over the land like Gulliver among the Lilliputians, his arms spread wide, as if to embrace his domain. On his head is a crown; in his left hand, a bishop's crosier, or staff; in his right, a sword to rule the land and defend it from all enemies. He dominates the land, and there is no question that it was he who brought it peace, order, and prosperity.

At a glance, the image seems like an advertisement for the virtues of a strong centralized monarchy on the French model. But there is something strange about this towering king. His body appears rugged, and he seems to be wearing some sort of scale armor. A closer inspection reveals the truth: they are not scales, but people. What appears to be the king's human body is in fact composed of the men of the commonwealth. Each single man is powerless, nothing but a minuscule component of this enormous body. But together, and working with a single will, they are the all-powerful Leviathan.

So overpowering and all-encompassing is the state depicted in the frontispiece that it leaves no room for any independent institutions. For

Hobbes, *any* institutions that are not directly dependent on the sovereign are a threat to the unity of the Leviathan and the stability of the state, a threat that, if not checked, will breed disagreements and conflicts and lead, once again, to civil war. The worst offender in Hobbes's book is the Catholic Church, which claimed ascendancy over all civic authorities and consequently earns itself an entire section in the *Leviathan*, entitled "The Kingdom of Darkness." In England, the rebellious Parliament is, of course, anathema to Hobbes, but so are seemingly tame institutions such as the Anglican Church and the universities of Oxford and Cambridge. Anglicanism is preferable to most churches because it is at least nominally subject to the king, even if, in Hobbes's opinion, Anglican clerics showed far too much independence. Other denominations, and especially Presbyterianism, are far worse, because they set up their own rule separate from the commonwealth, and Hobbes is not above blaming them directly for the onset of the civil war. The universities earn Hobbes's ire partly because they traced their intellectual roots to medieval Scholasticism, and so are tainted by association with the "Kingdom of Darkness." But more fundamentally, the universities seem to Hobbes dangerous breeding grounds for doctrines and ideas that might conflict with the will of the sovereign, leading to open-ended controversies. And controversy, the Earl of Newcastle, Hobbes's protector, warned Charles II some years after he was restored to the throne, "is a Civill Warr with the Pen, which pulls out the sorde soon afterwards."

Deciding which opinions and doctrines should be taught and which should be banned, in the universities and elsewhere, is the prerogative of the sovereign alone, Hobbes insists. If clergymen are allowed to preach what they want, and professors to teach what they want, then division, conflict, and civil war will soon follow. But Hobbes goes even further: the Leviathan decides not just which teachings are harmful to the state and which beneficial, but more fundamentally, what is right and what is wrong. In the state of nature, there is no right and wrong, according to Hobbes, since every man acts as best he can to secure his own interests. The notions of right and wrong arrive on the scene only with the Leviathan, and the standard is simple: "the law, which is the will and appetite of the state, is the measure," and nothing else. Following the law laid down by the sovereign is right, breaking it is wrong, and that's all there is to it. Anyone who appeals to other sources of authority, such as God,

The Leviathan. From the frontispiece to the 1651 edition of *Leviathan*. "Non est potestas super terram quae comparetur ei," declares the text surrounding the crown: "There is no power upon the Earth that compares to him."
(British Library Board / Robana / Art Resource, NY)

tradition, or ancient rights, is undermining the unity of the common-wealth, and likely plotting against it to benefit his own interests. Right and wrong, good and evil—all are in the hands of the sovereign.

As an exercise in political theory, *Leviathan* is as bold a tract as ever was, and rightly deserves its high place in the annals of Western thought. But does Hobbes's prescription, for all its brilliance, describe any actual historical state? There have been, no doubt, regimes that strove for the ideal, most infamously the mighty totalitarian states of the twentieth century, among them Hitler's Germany and Stalin's Soviet Union. These, too, were states in which the people were (in principle, at least) united in the person of their leader, whose will spoke for the entire nation. Like the Leviathan, they, too, tolerated no dissent, considering it an offense to the national will (in Germany) or the forward march of the proletariat (in Russia). But even these dark regimes of the last century never achieved what Hobbes had in mind. For one thing, they both had to deal with real dissent, however sporadic or muted, whereas Hobbes virtually defined dissent out of existence. On a deeper level, both the German Nazis and the Soviet Communists saw themselves as standing for some higher ideal, whether the mystical destiny of the German nation or the world revolution that would lead to a classless utopia. Their totalitarian states were a means to accomplish these higher goals.

The Hobbesian Leviathan is something else: No higher truths—religious, mystical, or ideological—exist outside the law laid down by the sovereign, and anyone who claims that there are such truths is designated an enemy of the state. The Leviathan is the be-all and end-all, and exists for itself alone. As a bulwark against chaos, it is absolutely necessary, but it stands for nothing, and steadfastly denies that any higher ideals or purposes even exist. All that counts are the state, the sovereign, and his will.

Hobbes's proposed solution to the crisis of the 1640s made his reputation as an original thinker, but it also put him at odds with his contemporaries—not only the Parliamentarians, whom he despised, but also the Royalists, whom he intended to support. There was, to be sure, much in *Leviathan* that appealed to the king's men, most particularly the insistence on a strong central government led (preferably) by one man, and the condemnation of Parliamentarians and all dissenters as traitors to the commonwealth pursuing their own selfish interests. But there was also much in the book to trouble dedicated Royalists. A legitimate king, after all, rules his lands and people by divine right; he is God's chosen, and can be replaced by no other. The Leviathan, in contrast, rules not by right, but as a practical necessity to prevent civil war. Anyone, in principle, can fill this role, as long as he is capable of defending the land and preserving the peace. If a king fails to do this, then he could in principle be replaced by someone else—as in fact happened in England. Some Royalists became so concerned about this that they accused Hobbes of having written *Leviathan* in support of Cromwell's Protectorate, not of the king. This was not true—*Leviathan* was published two years before Cromwell became lord protector—but it was not unreasonable: *Leviathan* is in truth not a defense of legitimate monarchy, but an argument for a totalitarian dictatorship. Add to that the fact that *Leviathan* managed to offend all clergymen, both Anglicans and the court in exile's French-Catholic hosts, and it becomes clear why this supposedly Royalist tract did not endear Hobbes to the king's followers. Instead of being celebrated as a Royalist standard-bearer, he was relieved of his tutoring duties, banished from court, and soon found himself, ironically, back in revolutionary England.

Much to Hobbes's disappointment, there seemed to be no takers for *Leviathan*'s extreme prescription for ending the civil war. But Hobbes knew he was right, regardless of what others thought. All that was needed to overcome their skepticism, he concluded, was to spell out his entire philosophy more clearly and comprehensively. Once he explained, step by step, precisely how he had arrived at his conclusions, the doubters, he was sure, would have no choice but to accept his political prescriptions. Now, Hobbes had never intended his political tracts *De cive* and *Leviathan* to stand alone. As conceived during his long, silent decades on the Cavendish estates, they were supposed to be the final part of a complete philosophy that would cover all facets of existence. The political crisis caused Hobbes to write the last part first and rush it to publication, in the hope (soon disappointed) that it would end the war and restore the king. Now that he was back in England, and working to buttress his case, he decided to go back and write the first two parts.

In 1658, Hobbes published *De homine*, a tract on human nature that confirmed his reputation for misanthropy but, compared to *Leviathan*, received little attention. His core principles, however, were laid down in

De corpore ("On Matter"), which came out three years earlier. *De corpore* is a dense, technical book aimed at professional philosophers and is concerned with such abstruse matters as whether universals truly exist and whether matter is nothing more than extension. It has none of the lively imagery or fiery rhetoric of *Leviathan*, and probably had only a fraction of its readership. Nevertheless, it was *De corpore*, not *Leviathan*, that embroiled Hobbes in a personal war that lasted for the rest of his life. For, even before the book saw light, a man determined to be his enemy was secretly obtaining the unpublished text from Hobbes's printer and preparing a devastating response. This man was not a rival philosopher waiting to challenge Hobbes on some technical definition of matter or motion. His name was John Wallis, and he was a mathematician.

7

Thomas Hobbes, Geometer

IN LOVE WITH GEOMETRY

Hobbes's introduction to mathematics is the stuff of legend. Having studied no mathematics at Oxford, he encountered it quite accidentally at age forty, while visiting Geneva with one of his young charges. "Being in a gentleman's library," his biographer Aubrey reports, "Euclid's *Elements* lay open, and 'twas the 47 El. Libri I" (i.e., theorem number 47 in the first book of *The Elements*). This, as anyone educated in classical mathematics knew, was the Pythagorean theorem, which states that the square of the hypotenuse of a right triangle is equal to the sum of the squares of its two legs. Hobbes read the proposition. "'By God' sayd he, 'this is impossible!' So he reads the demonstration of it, which referred him back to such a proposition," which in turn referred him to an earlier one, and so on, "that at last he was demonstratively convinced of that trueth." This, according to Aubrey, "made him in love with geometry."

In the years that followed, Hobbes worked hard to make up for his late start in the field. By the 1640s he was in regular contact with some of the leading mathematicians of the day, including Descartes, Roberval, and Fermat, and when the English geometer John Pell fell into dispute with his Danish colleague Longomontanus, he considered Hobbes enough of an authority to seek his public support. When Descartes died some years later, the French courtier Samuel Sorbière hailed him as one of the world's greatest mathematicians, along with "Roberval, Bonnel, Hobbes, and Fermat." Sorbière, it must be said, was Hobbes's good friend, and his high opinion of the Englishman's mathematical talents may not have been shared by all. His evaluation does nevertheless show that when Hobbes was made mathematical tutor to the exiled Prince of Wales, he was one of the most respected English mathematicians of the day.

Why was Hobbes so taken with mathematics? Aubrey's tale provides a crucial clue here: in geometry, each result is built on another, simpler one, so one can proceed, step upon logical step, starting with self-evident truths and moving to ever-more-complex ones. By the time a reader reaches some truly unexpected results, such as the Pythagorean theorem, he is "demonstratively convinced of that trueth." This, to Hobbes, was an amazing accomplishment: here was a science that could actually prove its results, leaving not a shadow of a doubt about their veracity. Consequently, he considered it "the only science it hath pleased God hitherto to bestow on Mankind," and the proper model for all other sciences. All sciences, he believed, should proceed like geometry, since "there can be no certainty of the last conclusion, without a certainty of all those affirmations and negations on which it was grounded and inferred." No field but geometry, Hobbes noted, has as yet achieved the required level of systematic certainty, but this was about to change: now that Hobbes had arrived upon the scene, he was ready to provide the true philosophy, which would be structured as systematically as geometry, and whose results would be just as certain.

Over the more than four centuries since his birth, Hobbes has been accused of many things: during his lifetime he was accused (probably falsely) of living a dissolute, immoral life and (probably correctly) of being an atheist and promoting irreligion. In later days he has been accused of an unjustifiably grim view of human nature, and of serving as inspiration to some of the most oppressive regimes in human history. But no one, in all this time, has ever accused Hobbes of excessive modesty. Indeed, when Hobbes came to present his new geometrically inspired philosophical system, there was not a trace of this lamentable trait. To the contrary, his texts simmer with brashness and provocation.

For most of its history, Hobbes explains in the dedication of *De corpore* to his Cavendish patron, the Earl of Devonshire, the world had known almost no philosophy. True, the ancients made great advances in geometry, and more recently there had been some important steps forward in natural philosophy, thanks to the work of Copernicus, Galileo, Kepler, and several others. As for the rest of philosophy, from Plato and Aristotle to the present, it was worse than useless. "There walked in old Greece a certain phantasm," he wrote, ". . . full within of fraud and filth, a little like philosophy," which some people mistook for the real thing. Instead of teaching the truth, this pseudo-philosophy taught people to disagree and dispute, and all just so that the supposed "philosophers" would be lavishly compensated. The worst of it was "school divinity," the medieval Aristotelianism taught at the universities. This, Hobbes charges, was a "pernicious philosophy that hath raised an infinite number of controversies . . . and from those controversies, wars." Hobbes calls this abomination "Empusa," the Greek monster with one leg of bronze, the other of an ass, which was a harbinger of ill fortune.

Hobbes was about to change all that. Natural philosophy, he explains, may be young, dating no further back than Copernicus, but civil philosophy was even younger, according to Hobbes, being "no older than my own book . . . *de Cive*." In that book, for the first time, he had used unchallengeable reasoning to prove that all authority in the state, whether religious or civic, must derive from the sovereign alone. Now, in *De corpore*, he would complete the job: he would set down the true philosophy that would replace the fake and pernicious ones, and finally conquer the monster Empusa. For Hobbes, his philosophy was not a contribution to an ongoing conversation that had lasted thousands of years: it was, rather, a philosophy to end all philosophies, the one and true doctrine that would put an end to all discussion and debate. His book, he wrote, may be short, but it is nothing short of great, "if men count well as great." His critics, of whom there were many, might dispute this, but Hobbes cared little for them. They were simply envious of the work, and he, after all, was not "striving to appease," he noted with perfect candor. The brilliance of *De corpore* would vanquish them: I will "revenge myself of envy by encreasing it," he announced, without a hint of irony.

How would Hobbes's new philosophy conquer Empusa? The answer was simple: through geometry. The reason that the so-called philosophers of the past had failed was that they relied on flawed and inconclusive methods of reasoning. They taught dispute instead of wisdom, Hobbes charges, and they "determine[d] every question according to their own fancies." As a result, instead of bringing peace and unanimity, they fostered strife and civil war. Geometry, in contrast, compelled agreement: "For who is so stupid as both to make a mistake in geometry and also to persist in it, when another detects his error to him?" Consequently, geometry produces peace rather than discord, and Hobbes's philosophy would follow its lead. *De corpore*, he explains in the dedication, was written for "the attentive readers versed in the demonstrations of mathematicians," and some parts of it were written "to geometricians alone." But the implications of the geometric method extended to all fields: "Physics, ethics, and politics, if they are well demonstrated, are no less certain than the pronouncements of mathematics." If one but follows the clear and indisputable method of geometrical reasoning, he will without trouble "fright and drive away this metaphysical Empusa."

In his own work, Hobbes believed he fully followed the geometrical example: his philosophy (once all its parts were published) begins with simple definitions in *De corpore*, just as Euclid's *Elements* begins with definitions and postulates. And just as *The Elements* moves from the simple and self-evident to the complex and surprising, so Hobbes's opus proceeds through its three sections: *De corpore* ("On Matter"), *De homine* ("On Man"), and *De cive* ("On the Citizen"). From a discussion of definitions (which he calls "names"), he proceeds to the nature of space, matter, magnitudes, motion, physics, astronomy, and so on. Finally, at the end of this long chain of reasoning, he reaches the most complex and the most urgent topic of all, the one that justifies the entire enterprise: the theory of the commonwealth. Certainly there were those who disputed whether he had succeeded in living up to the geometrical standard, but Hobbes paid them no mind. His systematic and careful reasoning from first definitions, he was convinced, ensured that his conclusions about the proper organization of the state were absolutely certain. As certain, in fact, as Euclid's Pythagorean theorem.

THE GEOMETRICAL STATE

If Hobbes's entire philosophy was structured as a grand geometrical edifice, this was particularly true of his political theory. This was because the commonwealth shared a fundamental feature with geometry: both were created entirely by humans, and therefore were fully and completely known to humans. "Geometry is . . . demonstrable, for the lines and figures from which we reason are drawn and described by ourselves; and civil philosophy is demonstrable, because we make the commonwealth ourselves." Our knowledge of how to create the ideal state is perfect, Hobbes claims, just like our knowledge of geometrical truth. In *Leviathan*, Hobbes put this principle into practice, creating what he believed was a perfectly logical political theory whose conclusions were in all respects as certain as geometrical theorems.

It wasn't just the broad principles of the commonwealth that possessed the certainty of geometrical demonstrations. The actual laws established by the Leviathan to govern the state also had the inescapable logical force of a geometrical theorem, and were as indisputably correct. As Hobbes puts it, "the skill of making and maintaining commonwealths consisteth in certain rules, as does arithmetic and geometry." This is because the laws themselves define what is right and true, and what is wrong and false. Before the commonwealth, in the state of nature, the terms *right* and *wrong* or *true* and *false* were empty words that referred to nothing. There was no justice or injustice, right or wrong, in the state of nature. But once men gave up their personal will to the great Leviathan, he laid down the law and gave the terms meaning: *right* is following the law; *wrong* is breaking it. Anyone charging that the decrees of the sovereign are "wrong" and should be changed is speaking nonsense, since what is "wrong" is defined by the decrees themselves. Opposing a law is as absurd as denying a geometrical definition.

Hobbes, of course, was far from the only early modern intellectual to idealize geometry. Only a few decades earlier, Clavius, too, had extolled the virtues of geometry, promising his fellow Jesuits that it would be a powerful weapon in the struggle against Protestantism. But apart from their admiration of geometry, Clavius and Hobbes had almost nothing in common. Clavius was a Jesuit scholar trained in Aristotelian philosophy and the methods of Scholastic disputation, which were embraced by his order and perfected at the Collegio Romano. He was also a Counter-Reformation warrior who fought to spread the word of God and bring about a Catholic spiritual awakening, and he abhorred Protestants, materialists, and heretics of all kinds. His life's ambition was to establish the Kingdom of God on Earth, which to him meant setting the Pope above all secular rulers, and the Church above all civic institutions. Hobbes, in contrast, had nothing but scorn for Scholastic disputation, believed that *spirit* was a meaningless term, and that only matter and motion existed in the world. The sole purpose of terms such as *spirit* and *immortal soul* was to allow unscrupulous and corrupt clergymen to frighten men and subject them to their will. Finally, the notion that the Pope would rule over kings was intolerable to Hobbes. Any infringement on the absolute power of the civil sovereign would lead to disagreements, divisions, and, inevitably, civil war.

Clavius died in 1612, long before Hobbes published his tracts and almost certainly without ever having heard the Englishman's name. But if he had had a chance to read any of Hobbes's works—*De cive, De corpore, Leviathan, De homine*—there is no question what his reaction would have been. To a devout Jesuit such as Clavius, Hobbes was a godless materialist and a heretic, an enemy of the Catholic Church whose books should be banned. If Hobbes had ever been so unfortunate as to fall into the hands of Clavius and his brethren, he would have been lucky to escape the stake. Meanwhile, Hobbes's verdict on the Jesuits was no less harsh: their goal, he argued, was to scare men and "fright them from obeying the laws of their country." This, as far as Hobbes was concerned, was true of all clergymen, but he reserved special scorn for the Catholic Church. The Jesuit dream of a universal and all-powerful Church, ruled by the Pope, was to Hobbes the darkest of nightmares.

Only on the role of geometry were these two natural enemies in perfect agreement. Euclidean geometry, Clavius believed, was a model of correct logical reasoning, which would ensure the triumph of the Roman Church and the establishment of a universal Christian kingdom on earth, with the Pope at its apex. Hobbes's Leviathan state was, in many ways, the precise opposite of the Jesuits' Christian kingdom: it was ruled by a civic magistrate who embodied the will of the people, not by the Pope, who derived his authority from God; its laws were derived from the Leviathan's will, not from divine or scriptural injunctions, and the

Leviathan would never tolerate any clerical infringement on his absolute powers. But in their deep structure, the Jesuit papal kingdom and the Hobbesian commonwealth are strikingly similar. Both are hierarchical, absolutist states where the will of the ruler, whether Pope or Leviathan, is the law. Both deny the legitimacy, or even the possibility, of dissent, and each assigns to each person a fixed and unalterable place in the order of the state. Finally, both rely on the same intellectual scaffolding to guarantee their fixed hierarchy and eternal stability: Euclidean geometry.

Today, Euclidean geometry is just one narrowly defined area of mathematics, albeit one with an extraordinarily long and impressive pedigree. Not only is it just one among many mathematical fields, it is also, since the nineteenth century, just one among an infinite number of geometries. It is taught to secondary school students today partly because of tradition and partly because it is thought to impart the powerful method of rigorous deductive reasoning. Beyond that, it is of little interest to practicing mathematicians. But things were very different in the early modern world, when Euclidean geometry was viewed by many as one of the towering achievements of humanity, the unassailable bastion of reason itself. To Clavius, Hobbes, and their contemporaries, it seemed natural that geometry would have implications far broader than for objects such as triangles and circles. As the science of reason, it should apply to any field in which chaos threatened to eclipse order: religion, politics, and society, all of which were in a state of profound disarray in this period. All that was needed was to use its methods in the afflicted fields, and peace and order would replace chaos and strife.

Euclidean geometry thus came to be associated with a particular form of social and political organization, which both Hobbes and the Jesuits strived for: rigid, unchanging, hierarchical, and encompassing all aspects of life. To us, who can look back on the rise and fall of bloody totalitarian regimes in recent centuries, it is a chilling, repellent vision. But at the dawn of the modern age, with the old medieval world in shambles and nothing to replace it, perspectives were different. To Clavius, to Hobbes, and to many others, it seemed that the answer to uncertainty and chaos was absolute certainty and eternal order. And the key to both, they believed, was geometry.

THE PROBLEM THAT WOULD NOT BE SOLVED

Beautiful and powerful as it was, Euclidean geometry was not free from flaws, as Hobbes discovered to his dismay when he began studying the subject in depth in the years after his encounter with the Pythagorean theorem. The difficulty was that certain classical problems of mathematics, known since ancient times, still defied solution: the squaring of the circle, the trisection of an angle, the doubling of the cube. Despite the efforts of the greatest mathematicians over a span of nearly two millennia, these classical conundrums still defeated every effort to solve them.

This was very bad news for Hobbes's science of politics. If geometry is fully known, as he declared it must be, then it should have no unsolved, not to mention insoluble, problems. The fact that it does suggests that it possesses dark corners where the light of reason does not shine. And if geometry, which deals with simple points and lines, is not fully understood, how can one expect the theory of the commonwealth, which deals with the thoughts and passions of men, to be perfectly known? If geometry has its blind spots, then the science of politics may well have some, too, and they are likely to be far greater and more significant than those of geometry. For Hobbes, as long as the classic geometrical problems remained unsolved, his entire philosophical edifice remained insecure, and the Leviathan state a political house built on sand.

In order to secure the foundations of his political theory, Hobbes set out to solve the three classical outstanding problems of geometry. Initially he seems to have believed that this should not be too difficult. Surely, he thought, just as he had corrected the errors of all past philosophers, he could also correct the errors of all past geometers. And he can perhaps be excused for his unwarranted optimism, because part of the reason the problems had attracted the attention of the greatest mathematicians over centuries was that they were easily stated, and appeared deceptively simple. "The quadrature of the circle" means constructing a square equal in area to a given circle; "the trisection of the angle" means dividing any given angle into three equal parts; and "the doubling of the cube" means constructing a cube of double the volume of a given cube. How hard could it be to solve such questions? As it turns out, very hard. In fact, impossible.

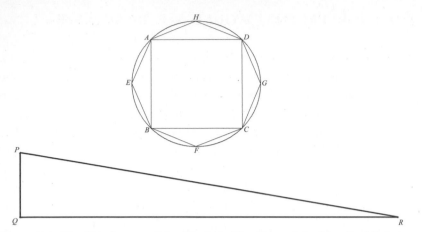

Figure 7.1. The Quadrature of the Circle 1. The Case of the Inscribed Polygon.

To understand why this is so, consider the problem that most interested Hobbes and to which he devoted an entire chapter in *De corpore*: the quadrature of the circle. Already in the second century BCE, the polymath Archimedes of Syracuse proved that the area enclosed in a circle is equal to that of a right triangle whose two perpendicular legs are equal to the radius and to the circumference of a circle—that is, *PQ* and *QR* in Figure 7.1. Archimedes proved this result by looking at polygons inscribing and circumscribing a circle. The greater the number of sides of a polygon, the closer its enclosed area is to that of the circle. Now, Archimedes reasoned, let us consider the circumscribed octagon *AHDGCFBE*. Its area is equal to that of a right triangle whose two perpendicular sides are equal to the apothem and the sum of the octagon's sides (the "apothem" being a vertical line from the center of the polygon to its side). This is obvious if we think of the area of the octagon to be the sum of eight triangles with bases on the sides of the octagon and the apex at the center. The area of each triangle is half its base times the apothem, and consequently the area of the entire octagon is half all the bases together times the apothem—that is, the area of the right triangle in question.

Let us look, Archimedes continued, at the area of the circle, which we call *C*, and the area of the right triangle whose perpendicular sides are its radius and circumference, which we call *T*, and the area of an inscribed polygon with n sides, which we call I_n. Let us assume for the moment that the circle is greater than the triangle—that is, $C > T$. Archimedes had already shown that as we increase the number of sides of the inscribed polygon, it approaches the area of the circle as closely as we please. Consequently, there is a number of sides n for which the area of the polygon falls between the area of the circle and the triangle that is greater than the triangle but (being inscribed) still smaller than the circle. In modern notation: $I_n > T$.

This, however, is impossible, because the area of I_n is equal to a right triangle whose perpendicular sides are its apothem and the sum of its sides. The apothem is less than the radius of the circle, and the sides are less than the circle's circumference. This means that $I_n < T$, contradicting the original assumption, and the triangle cannot be smaller than the area of the circle. Archimedes then assumed that the triangle is larger than the area of the circle, and repeated the argument, looking this time at the circle's circumscribing polygon. Once again he showed that as we increased the number of the polygon's sides, it approached the area of the circle as closely as one wished. But since it always remained greater than the triangle T, the area of the triangle could not be greater than the circle, contradicting the assumption. Consequently, the only remaining possibility is that the triangle is equal to the circle, or $C = T$. QED.

Archimedes's demonstration is an example of the classical "method of exhaustion," and it is a perfectly rigorous Euclidean proof. From this it might appear that the quadrature of the circle has been accomplished: the triangle is equal in area to the circle, and it is a simple matter to construct a square whose area is equal to a triangle. Problem solved? Not by the standards of the classical geometers. Archimedes did indeed demonstrate that the area of a circle is equal to the area of a particular triangle, but he did not "construct" the triangle with compass and straightedge, the only tools allowed in Euclidean constructions. For a quadrature to be acceptable to classical geometers, one would have to begin with a given circle and, through a finite series of steps, using only compass and straightedge, produce the required triangle. Archimedes didn't do this: he proved that the area of the circle is equal to that of the triangle, but he did not show how to construct a triangle with such measurements from a circle. Hence his proof, elegant though it was, and correct though it was, was not a quadrature.

To us, these rigid standards of classical geometry might seem rather fussy, if not pointless. Modern mathematicians do not limit themselves to constructive proofs, not to mention constructive proofs by straight-edge and compass alone. Indeed, Archimedes's proof is more than satis-factory for anyone who wants to find out the area enclosed in a circle. But Hobbes thought differently. To him, the fact that geometry was built up step by step from its simplest components to ever-more-complex re-sults was what made it an appropriate model for philosophy and for the science of politics. To be worthy of this status, it is essential that geom-etry itself not stray from this model, and that it always construct its ob-jects by proceeding systematically from the simple to the complex, using the most rudimentary tools. To Hobbes, the classical standards were not arbitrary impositions, but the very heart of what geometry is and should be. To resolve the quadrature of the circle, it is therefore necessary to construct a square with the area of a circle, as the ancient geometers required, and as Archimedes failed to do.

LEVIATHAN SQUARES THE CIRCLE

Why had mathematicians failed to square the circle despite repeated ef-forts over thousands of years? Quite a few mathematicians in Hobbes's day began to suspect that the reason was that the three classical problems were simply insoluble, but that is a possibility that Hobbes could never entertain. If it was to serve as the keystone of his philosophy, geometry had to be perfectly knowable, and there could therefore be only one an-swer: the mathematicians were working from flawed assumptions. Once correct assumptions were put in place, true results would grow naturally from them, for "it is in the sciences as it is in plants," Hobbes wrote: "growth and branching is but the generation of the root continued."

According to Hobbes, the problem with Euclid—or rather, as he pre-ferred, with Euclid's followers and interpreters—is that the definitions they use are overly abstract and refer to nothing in the world. The Euclidean definition for a point, for example, is "that which has no parts," the defini-tion of a line is "breadthless length," and a surface is "that which has breadth and length only." But what do those definitions mean? "That which has no parts," Hobbes argues, "is no Quantity; and if a point be not quantity . . . it is nothing. And if *Euclide* had meant it so in his definition . . . he might have defined it more briefly (but ridiculously) thus, *a Point is nothing.*" Precisely the same is true for the definitions of a line, a surface, or a solid: they have no referent, and are consequently meaningless.

Only one type of definition would satisfy Hobbes: one based on matter in motion. In fact, as a materialist to the core, Hobbes believed that there was nothing in the world *except* matter in motion. All the fancy talk of abstractions and immaterial spirits was merely a ploy to gain power over men. Points, lines, and solids, the building blocks of all geometry, must therefore be defined in terms of things that actually exist:

> If the magnitude of a body which is moved (although it must always have some) is considered to be none [*nulla*], the path by which it travels is called a *line* or one simple dimension, and the space it travels along a *length*, and the body itself is called a *point*.

Surfaces and solids are then defined in the same way: a surface by the motion of a line, a solid by the motion of a surface.

The odd thing about Hobbes's definition is that in his scheme, points, lines, and surfaces are actual bodies, and therefore have magnitude. Points have a size, lines have a width, and surfaces have a thickness. This was heresy for traditional geometers, who from the time of Plato (and likely before) viewed geometrical objects as pure abstractions whose crude physical manifestations were but pale shadows of their true perfection. And if the philosophical issue of the true nature of geometrical objects was not reason enough to reject Hobbes's approach, there were also the practical questions of how to account for these strange magnitudes in geometrical proofs. What width must one assign to a line, or thickness to a plane? And do traditional proofs hold true for such unorthodox objects? There were no good answers to these questions, and it is not surprising that, to traditional mathematicians, the idea of treating geometrical objects like material bodies with width and breadth seemed like the end of geometry.

Aware of these difficulties, Hobbes argued that although geometrical objects, being bodies, do have positive magnitudes, they are considered in proofs without regard to their dimensions. That is, a point is a body that "is considered" to have a zero size, a line is a path that "is considered" to have length but zero width, and so on, even though, in truth, points have size and lines have width. What exactly Hobbes meant by this argument is far from clear. He seems to be trying to balance his insistence that everything, including geometrical bodies, is made of matter in motion with the traditional demands of Euclidean geometry, which he greatly admired. What is clear is that orthodox geometers were far from convinced.

Conceiving geometrical objects as material bodies was one key component of Hobbes's geometry. The other was another seemingly physical attribute: motion. Lines, surfaces, and solids were all created by the movement of bodies, and Hobbes's geometry accounts for this. The most minuscule possible motion "through a space and time less than any given," he called "conatus"; the speed of the conatus he called "impetus." To account for how these minuscule motions added up to complete lines and surfaces, he drew on a surprising source: Cavalieri's indivisibles.

Hobbes, in fact, knew Cavalieri's work better than almost any mathematician in Europe. He was one of the few who had actually read Cavalieri's dense tomes, and did not rely on Torricelli's later adaptation. But how could someone so insistent on the logical clarity of geometry as Hobbes adopt the notoriously murky indivisibles so often attacked for being logically inconsistent and paradoxical? The answer lay in Hobbes's unconventional interpretation of indivisibles. Cavalieri's indivisibles, according to Hobbes, were material objects with a positive magnitude: lines were in fact tiny parallelograms, and surfaces were solids with a minuscule thickness, which for the sake of calculation was "considered" zero. As Hobbes's friend Sorbière explained it, "instead of saying that a line is long, and not broad, [Hobbes] allows of some very little breadth, of no matter of Account, except it be for a very few Occasions." These points, lines, and surfaces were not fixed, stationary objects in Hobbes's geometry: like other physical bodies, they could and did move. With a given conatus and impetus, points generated lines, lines generated surfaces, and surfaces generated solids.

Clavius, the champion of classical geometry, would have been appalled at Hobbes's unconventional geometry. For him, lines with breadth and surfaces with thickness, moving through space with a given impetus, had no place in the pure, immaterial realm of geometrical objects.

But Hobbes was not trying to overthrow traditional geometry. To the contrary, he was trying to reform it by founding it on the principles of matter in motion, thereby making it even more rigorous and more powerful. "Every demonstration is flawed," he argued, if it does not construct figures by the drawing of lines, and "every drawing of a line is a motion." No one had ever seen a point without size or a line with no width, and it was obvious that such objects did not and could not exist in the world. A true, rigorous, and rational geometry must be a material geometry, and that is what Hobbes created. The new material geometry, Hobbes was convinced, would easily resolve all outstanding problems (such as the quadrature of the circle) that had vexed geometers for millennia. It would be what traditional geometry aspired to be: a perfectly knowable system.

Unfortunately for Hobbes, his effort at squaring the circle did not proceed as smoothly and naturally as a plant growing from its roots. By the early 1650s he was letting it be known among his friends that he had succeeded in squaring the circle, but although he was very proud of his accomplishment, he had no immediate plans for publishing it. He was too preoccupied, it seems, with preparing *De corpore* for publication. But in 1654, Hobbes received a challenge: Seth Ward, an old acquaintance who was now the Savilian Professor of Astronomy at Oxford, anonymously published a detailed defense of the universities against Hobbes, who had dismissed them in *Leviathan* as servants of the "Kingdom of Darkness." Noting that he "hath heard that Mr. *Hobbs* hath given out that he hath found the solution of some problemes, amounting to no less than the quadrature of the circle," Ward promised to "fall in with those who speake loudest in his praise" if Hobbes published a true solution.

It was a trap, and Hobbes knew it. Ward, he realized, was trying to provoke him into disclosing his proof, convinced that Hobbes could not possibly have solved a problem that had stood for millennia. Nevertheless, confident that his reformed geometry would succeed where Euclid had failed, Hobbes took the bait. He quickly added a chapter to *De corpore* that included a proof of the classical problem. By squaring the circle, Hobbes believed he would all at once embarrass his self-satisfied detractor, demonstrate the superiority of his refurbished geometry over the traditional one, and by extension establish the truth of his philosophical system and political program. Despite the risks, it was an opportunity he could not turn down.

But Hobbes's plan got off to a bad start. After sending the manuscript of *De corpore* out to a printer, with his quadrature of the circle in chapter 20 included, he had second thoughts. Was his proof really as unassailable as he had thought? He showed it to some trusted friends, and they quickly pointed out his error; he rushed a correction to the printer. He probably would have wanted to remove the proof from the published book entirely, but it was too late for that, so he came up with an ingenious solution. As was customary in books of the period, the top of each chapter included a list of its contents, so Hobbes decided to leave the proof in place but change its description in the list: instead of calling it a quadrature of the circle, he now entitled it "from a false hypothesis, a false quadrature." This may have saved him from the embarrassment of putting his name to a demonstrably false proof, but it also nullified the value of the demonstration. To compensate, in the same chapter, he added a second proof, but this turned out, on closer inspection, to be a mere approximation. He admitted as much in the title he assigned to it at the top of the chapter, and moved on. A third proof fared no better: although he confidently called it "Quadratura circuli vera" at the top of the chapter ("A True Quadrature of the Circle"), he was eventually forced to add a remarkable disclaimer at the end of the chapter:

> Since (after it was written) I have come to think that there are some things that could be objected against this quadrature, it seems better to warn the reader of this than to delay the edition any further . . . But the reader should take those things that are said to be found exactly of the dimension of the circle and of angles as instead said problematically.

Problematically indeed. In one single chapter of *De corpore*, despite his confident assertions to his friends and his bravado in taking up Ward's challenge, Hobbes had failed three times to square the circle. Instead of an incontrovertible proof of the quadrature of the circle, he had produced one "false quadrature," one approximation, and one proof that should be taken "problematically." This is hardly the result Hobbes had hoped for when he set out to create a logically irrefutable geometry and, from that, a logically irrefutable philosophy. Instead, he was left with imprecise and questionable results that, rather than establishing a new and peaceful geometric regime, only invited more controversy and speculation.

If this was not bad enough, yet another new enemy stood ready to make the most of Hobbes's discomfiture and turn it into public humiliation. John Wallis, the Savilian Professor of Geometry at Oxford, was as concerned as his colleague Ward about the dangerous influence of the man they called "the Monster of Malmesbury." Closely following Hobbes's plans for *De corpore*, Wallis used his connections to obtain the unpublished sheets of the book from the London printer. It was an underhanded and perhaps immoral tactic, but it proved extremely effective at undermining Hobbes's mathematical credibility. The sheets gave Wallis a head start on his rebuttal of *De corpore*'s mathematical claims, which he published mere months after Hobbes's book appeared in April 1655. But Wallis went even further: by comparing the unpublished with the published version of the text, he was able to reconstruct the entire chain of events that had led to Hobbes's odd and strangely contradictory claims in chapter 20. Hobbes's confident claims of success repeatedly followed by embarrassing retractions and qualifications were all gleefully exposed in Wallis's *Elenchus geometriae Hobbianae*. Hobbes's reputation as a leading mathematician never recovered.

THE HOPELESS QUEST

As it happens, Hobbes's reform of the foundations of geometry made little difference for the squaring of the circle. Although his friend Sorbière confidently declared that Hobbes's insistence on the material nature of points and lines finally provided the "Solutions of Problems that have hitherto remained insoluble, such as the squaring of the Circle and the doubling of the Cube," experience proved otherwise. The attempted proofs of the quadrature in *De corpore* were indeed unorthodox, relying on the motion of points and lines to produce the lines and surfaces, but in the end they were no more conclusive than the efforts of classical geometers. Whether done by traditional Euclidean methods or by Hobbes's new geometry, the task was hopeless: it is simply impossible to construct a square that has the same area as a given circle using only a straightedge and compass. Hobbes could not accept this, because it would mean that geometry harbored unknowable secrets, but quite a few mathematicians of his generation, including Wallis, suspected that it could not be done. At the very least, the mere fact that geometers had been trying and failing the challenge for nearly two millennia suggested that squaring the circle was not a good use of a geometer's time. The proof that the quadrature of the circle is impossible, however, had to wait two more centuries, and relied on a kind of mathematics that neither Hobbes nor Wallis could imagine.

To get an idea of why squaring the circle is a hopeless task, consider a circle with radius r. As every mathematics student today knows, the area of such a circle, in modern notation, is πr^2. Consequently, the side of a square whose area is equal to a circle is $\sqrt{\pi r^2}$, or, more simply, $\sqrt{\pi} r$. The magnitude of r was given in the problem, and we can assume for the sake of convenience that it is 1. All that remains is to construct a line with the length $\sqrt{\pi}$, and since Euclid shows how to construct a line that is the square root of another, this means constructing a line of length π, using only compass and straightedge. And that, as it turns out, is impossible. The reason, as eighteenth-century mathematicians discovered, is that classical geometrical constructions can produce only algebraic magnitudes—that is, magnitudes that are roots of some algebraic equation with rational coefficients. It took another century, but in 1882, the mathematician Ferdinand von Lindemann proved that π is not an "algebraic" number, but rather a new kind of number he called "transcendental," because it is not the root of any algebraic equation. Consequently, a line of length π cannot be constructed by compass and straightedge, and the squaring of the circle is impossible.

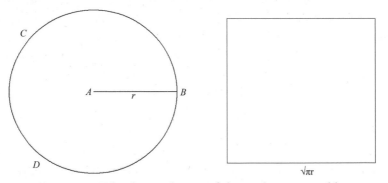

Figure 7.2. Why the quadrature of the circle is impossible.

All this, however, was centuries away when Hobbes published his quadratures of the circle. He knew nothing of algebraic and transcendental numbers or the limitations of classical constructions, not to mention of Lindemann's proof, and remained convinced throughout his life that his method was bound to lead to a true quadrature of the circle. His missteps in the first edition of *De corpore* he attributed to overhastiness, and he went on to supply corrected proofs in subsequent editions of the work, as well as in other treatises. Wallis stalked his steps, supplying refutations of each and every proof he offered, and other leading mathematicians joined him. Initially Hobbes reluctantly conceded the mathematical criticisms of his work, which led him to revise his proofs again and again. In time, however, he lost patience with his band of critics: he grew less and less receptive to their arguments, dismissing them as the work of small and envious minds that refused to acknowledge his profound contributions to geometry. Pedantry, prejudice, and pettiness were, for Hobbes, the only possible explanations for the mathematical community's hostility to his accomplishments. That his path was the true one, he did not doubt.

Hobbes never retreated from his unshakeable conviction that he had squared the circle. A few years before his death at the grand old age of ninety-one, he handed Aubrey a short autobiography with a list of his life's accomplishments. Among these, mathematics took pride of place. Hobbes took credit for having "corrected some principles of geometry" and for having "solved some most difficult problems, which had been sought in vain by the diligent scrutiny of the greatest geometers since the beginning of geometry." He then went on to list seven major problems he had solved, including the calculation of centers of gravity and the division of an angle. But there was no doubt of which "accomplishment" Hobbes was proudest: the very first item on the list was the quadrature of the circle.

8

Who Was John Wallis?

THE EDUCATION OF A YOUNG PURITAN

In 1643, while Hobbes was in Paris navigating the political maze of a court in exile and perfecting his philosophical system, a young clergyman in London was also trying his hand at philosophizing. How do we know what we know, and how can we be certain that what we know is true? he asked. The questions may have been similar to those Hobbes was asking at around the same time, but the answers were not. "A *Speculative* knowledge," the clergyman wrote, in a short booklet he called *Truth Tried*, "is found even in the Devils" in exactly the same measure as it is found in "the Saints on Earth." This, he explained, is because even devils are rational creatures, and can follow a logical argument as well as the children of God. There is, however, a higher form of knowledge: "*Experimentall* knowledge," which is of altogether "another nature." With this kind of knowledge "we do not only *Know* that it is so, but we *Tast* and *See* it to be so." Unlike beliefs based on speculation, "truths thus cleerly and *sensibly* . . . reveiled to the soul, it seems not in the power of the Will to reject."

The young clergyman was none other than John Wallis, then only twenty-seven years old and but a few years removed from his university days at Cambridge. It is very probable that he had never heard of Thomas Hobbes, the man who would become his obsession in later years but who was then just an obscure Cavendish family retainer with philosophical pretensions. It goes without saying that Hobbes had never heard of Wallis. Nevertheless, years before they launched their bitter war that would last a quarter century, the stark contrast between them was already in evidence. Hobbes maintained that true knowledge began with proper definitions and proceeded by strict, logical reasoning; Wallis believed such knowledge belonged to the Devil as much as to God. Wallis held that the highest form of knowledge is based on the senses—on "seeing" and even "tasting" the truth; Hobbes scorned such sensual knowledge, considering it unreliable and prone to error. Only on one issue did the two seem to be in complete agreement: that mathematics is the science of correct reasoning and certain knowledge, and should serve as a model for all fields.

Hobbes's interest in mathematics is hardly surprising, as it was at the core of his philosophical and political system. In 1643 he was already a geometer of some repute, and his mathematical star would continue to shine for some time. But Wallis, in the same year, was not a mathematician at all, but rather a rising Presbyterian clergyman deeply engaged in parliamentary efforts to reform the Church. It was not a calling that required deep mathematical knowledge. What is more, his comments in *Truth Tried* do not suggest great admiration for mathematics' signature brand of reasoning. If the most certain knowledge is experimental knowledge, where does that leave mathematics? Certainly, the views Wallis expressed at age twenty-seven do not seem a promising starting point for a career in mathematics. Yet it would be only a few years before Wallis was appointed to one of the most prestigious mathematical chairs in all Europe, and not long thereafter that he would justify the appointment by proving himself one of the most creative and widely admired mathematicians in the world.

Why would a man of Wallis's vocation and beliefs devote his life to mathematics? It seems an odd and unlikely career choice, but he had his reasons, and as was the case with Hobbes, these reasons extended to his philosophical and political convictions. As with Hobbes, Wallis's political attitudes reflected a strong reaction to the chaotic years of the Interregnum, but the conclusions he drew from that time were very different. Whereas for Hobbes the only answer to the crisis was a dictatorial Leviathan state, Wallis believed in a state that would allow for a plurality of views and wide scope for dissent. And whereas Hobbes relied on the rigid edifice of Euclidean geometry to buttress his inflexible Leviathan state, Wallis relied on a novel mathematical approach, as flexible and powerful as it was paradoxical and controversial: the mathematics of the infinitely small.

Born in 1616, Wallis was a full generation younger than Hobbes, but his background was otherwise strikingly similar to that of his great antagonist. He, too, was a southerner, from the town of Ashford, in Kent, east of Hobbes's home village of Westport, in Wiltshire. Wallis's father, too, was a minister, though seemingly of a more respectable sort. Whereas the elder Hobbes was known more for his gambling than his learning, John Wallis Sr. was, according to his son at least, "a Pious, Prudent, Learned, and Orthodox Divine," as well as a graduate of Trinity College, Cambridge. As Wallis recalled in an autobiography he wrote when in his eighties, his father was a leader in the community and took his position seriously: "Beside his constant preaching twice on the Lords-day, and other occasional Sermons, and catechizing; he . . . did maintain a week-day Lecture, on Saturday. Which was much frequented . . . by very many of the Neighbour Ministers, the Justices of the Peace, and others of the Gentry."

Sadly for young John, his father died when he was only six years old, a fate that is also reminiscent of Hobbes's early years. But the Wallis family was apparently better off, and whereas Hobbes was sent to live with his uncle, Wallis's mother, Joanna, was able to keep her family together and see to the upbringing of her five children. And although she had many opportunities for remarrying well, she remained a widow after her husband's death "for the good of her Children," as Wallis put it years later. John was the third of Joanna's children, but the oldest son, and she took charge of his education with ardor. To make sure he had the benefit of the best instructors, she sent him to school, first in Ashford and later in the nearby town of Tenterden, where he studied English grammar and Latin. Even as a child, he wrote later, he was never content simply to "know" but always sought to understand as well: "For it was always my affectation, in all pieces of Learning or Knowledge, not merely to learn by rote, which is soon forgotten, but inform the grounds or reasons of what I learn."

In Christmas of 1630, when Wallis was thirteen, he transferred to Martin Holbeach's school in Felsted, Essex, a move that would do much to shape his future. Holbeach was no mere schoolmaster, but a famous Puritan minister, active in efforts to reform the Church government and often in open conflict with the archbishops and bishops who ruled the Church of England. In later years Holbeach became an ardent supporter of Parliament in its struggle with the king, and ultimately an advocate of independent church government, in opposition to both Anglicans and Presbyterians. His reputation as a godly man and reliable Parliamentarian, as well as an excellent scholar and teacher, was such that leading Puritans from around the country sent their sons to study under him. Among them was Oliver Cromwell, whose four sons were students at the Felsted school.

Wallis did not have the benefit of an illustrious ancestry, but he nevertheless captured the schoolmaster's attention with his quick mind and studious habits. "*Mr Holbech* was very kind to me," he recalled in his autobiography, "and used to say I came to him the best grounded of any Scholar that he received from another school." Under Holbeach's direction, Wallis improved his Latin and Greek, studied logic, and acquired a smattering of Hebrew, all of which served him well when he went on to university. But the schoolmaster's true influence went beyond academic instruction. Holbeach, bitter Royalists claimed in later years, "scarce bred any man that was loyall to his Prince," and Wallis was no exception to this rule. At Felsted he was drawn into the circle of Puritan divines opposed to the Anglican hierarchy, and learned to stand up for the rights of freeborn Englishmen in the face of perceived royal oppression. When civil war broke out a decade later, Wallis remained true to what he learned at Felsted, and unhesitatingly threw in his lot with Parliament.

There was one subject, however, that Wallis did not learn at Felsted, or at any of his other schools. At that time, Wallis explained, mathematics "were scarce looked upon as *Accademical* studies, but rather *Mechanical*; as the business of *Traders, Merchants, Seamen, Carpenters, Surveyors of Lands*, or the like." It simply was not considered part of a young gentleman's proper education, and therefore was not included in the curriculum of any of his schools. As a result, Wallis's first encounter with mathematics, much like that of Hobbes, was purely accidental. In December of 1631, Wallis was home in Ashford for the Christmas holiday when he noticed one of his younger brothers engaged in a peculiar activity. The younger boy was apprenticed to a tradesman in town, who was teaching him arithmetic and accounting to assist with the business. Wallis was curious, and the boy, flattered no doubt by his older brother's attentions, volunteered to teach him what he had learned. The two spent the rest of the holiday together going over the lessons, with the elder Wallis learning the basic tricks of accounting. "This was my first insight into *Mathematicks*," he noted wistfully years later, "and all the *Teaching* I had." For all mathematical knowledge beyond these early lessons with a younger brother, Wallis relied exclusively on himself.

Although both Wallis and Hobbes, by their own testimony, discovered mathematics by mere chance, the tales they tell of these moments are very different. Hobbes stumbled upon mathematics in a gentleman's library, when he was traveling on the Continent with his highborn companion. It was in an aristocratic milieu, one in which mathematics was part of classical learning, studied not for its utility but as one of the refinements of upper-class life. Wallis, in contrast, discovered mathematics in his mother's crowded and no doubt noisy house packed with brothers and sisters during Christmas break. There was nothing refined about it, but quite the opposite—it was considered fit for his apprentice younger brother, not for an aspiring young gentleman like himself.

Not only the environment, but also the kind of mathematics each man encountered was radically different. In the gentleman's library, Hobbes discovered majestic Euclidean geometry and was captivated by its cold and rigorous beauty. But the mathematics Wallis found contained no geometry at all, only the arithmetic and rudimentary algebra of bookkeeping. There were no theorems and no proofs in this kind of math, and not a whiff of the grand philosophical claims made on behalf of Euclidean geometry. Hobbes's geometry was for him an example of universal, unchallengeable truth, whereas Wallis's mathematics was just a pragmatic tool for tradesmen, mariners, and land surveyors to solve the problems that came their way. There was nothing in it of the imposing edifice that Hobbes so admired.

Both Wallis and Hobbes recounted these stories many years later, and their mature and well-considered views on mathematics may have colored their memories of their earliest experiences with the field. Regardless, there is no denying that the accounts capture something fundamental about the two men's contrasting approach to the discipline. For

Hobbes, mathematics remained a refined, aristocratic science admired for its strict logical inferences. Whether it was useful or not mattered not at all. For Wallis, there was nothing aristocratic about mathematics, and it remained at its core a practical tool for obtaining useful results, just as it was for the tradesman who first taught it to his brother. Whether it was logically rigorous or not was of no consequence.

Wallis was very pleased with what he learned from his younger brother during that fateful Christmas, and also with his own surprising talent for mathematics. From that time onward he continued studying it on his own, though "not as a formal Study, but as a pleasing Diversion at spare hours." He had no teacher and no guidance, but he practiced his skills as often as he could and read mathematical works that fell into his hands. He never imagined, however, that this leisurely entertainment would end up being his life's work. His career path lay elsewhere: as a serious and pious young man, eager to make his mark in the world, he was intent on becoming a minister like his father, and preaching the word of God. To do so, he first had to obtain a university degree from Oxford or Cambridge, which, at the time (and for several centuries thereafter), were primarily schools for the training of clergymen.

Wallis entered Cambridge in Christmas of 1632 and was admitted into Emmanuel College. The choice of college was likely not a coincidence. Emmanuel was known as Cambridge's "Puritan" college, founded specifically for the purpose of training Puritan divines. It was a natural destination for the Felsted-trained Wallis, and his mentor Holbeach may have used his connections among the Puritan clergy to secure Wallis's admission. Much of the university curriculum consisted of the brand of medieval Aristotelianism known as Scholasticism, and students were judged on their ability to contest or defend the teachings of ancient and medieval philosophers in public disputations. Hobbes had despised Scholasticism in his university days, and later condemned it in *Leviathan* as "Aristotelity," but Wallis did not find the practice nearly as odious. He quickly mastered the intricacies of syllogisms, he proudly reported in his autobiography, to the point where he was "able to hold pace with those who were some years my Seniors," and soon acquired a reputation as a "good Disputant." In this manner he learned not only logic, but also the other fields of the Aristotelian canon, including ethics, physics, and metaphysics. His focus was always on theology, however, and at Emmanuel there was much on offer. Building on what he already knew from his religious upbringing, he began studying systematic academic theology, and soon became proficient in it.

But while adhering closely to the required Aristotelian curriculum, Wallis did not fail to notice that there were also other intellectual winds blowing through the university in the 1630s. The geographical discoveries of the past century revealed entire continents that went unmentioned in the classical sources, a development that seriously undermined confidence in the authority of the traditional canon, and made geography one of the most vibrant fields of the age. Medicine had also moved well beyond the writings of Galen that were studied at the university, thanks to, among others, the anatomical atlas of Vesalius and the recent discovery of the circulation of the blood by William Harvey. No field, however, had produced more brilliant discoveries than astronomy. Ever since Copernicus first published his *De revolutionibus* in 1540, his theory that the earth revolved around the sun had steadily gained adherents, evolving from a far-fetched hypothesis into a widely accepted account of the structure of the heavens. The growing appeal of Copernicanism was helped along by Kepler's accurate calculation of the planets' true orbits and the relations between them, and by Galileo's stunning telescopic discoveries, described in *The Starry Messenger*. That Galileo was persecuted by the Roman Church for advocating the Copernican system, in the very years Wallis was attending Cambridge, only enhanced the theory's popularity in Protestant England. And it was no less an eminence than Sir Francis Bacon, former Lord Chancellor of England, who combined these diverse discoveries into a philosophical system that, he promised, would revolutionize human knowledge and man's power over nature.

These early stirrings of the scientific revolution were known at the time as the New Philosophy, a term that combined a certain glamour and great promise with a whiff of danger and unorthodoxy. None of it was included in the universities' ossified curricula, but that does not mean that it did not reach Oxford and Cambridge. The excitement of the new science was in the air, and professors and students alike gathered informally to study it. Wallis was among them, devoting time to the study of astronomy, geography, and medicine, in addition to pursuing his long-standing interest in mathematics. He even went so far as to defend the circulation of the blood in a public disputation—the first student at the university to do so. Those activities, which came in addition to his formal studies, likely gobbled up most of Wallis's waking hours during his university years. But Wallis possessed an insatiable intellectual curiosity, and a remarkable capacity for hard work. He figured, as he wrote later, that "Knowledge is no Burthen," and that if it will not, in the end, prove useful, it certainly won't hurt. It would not be many years before this side interest became the center of his activities, when he joined with others to found one of the first scientific academies in the world, the Royal Society of London.

Wallis earned his bachelor's degree in 1637 and his master's in 1640, and would have become a fellow at Emmanuel were it not for the fact that the college already had a fellow from Kent, and the statutes allowed for only one from each county. He instead became a fellow at Queens College, but left soon after, following his marriage. For the next few years he served as minister to a succession of London churches and as personal chaplain to several aristocrats who sided with Parliament in its struggle with the king. One of them was Lady Mary Vere, widow of the soldier and continental campaigner Sir Horatio Vere. One night, while Wallis was seated at table in Lady Vere's London home, a fellow chaplain brought in a letter written in code that had been intercepted by Parliamentary forces and jokingly asked he if he could decipher it. He took on the challenge, and much to his colleague's surprise, succeeded after only two hours. This first code was a simple one, but the art of deciphering hardly existed at the time, and the feat gave Wallis a reputation as something of a miracle worker. Later codes he encountered proved more complex, but Wallis enjoyed the challenge and managed to break a good number of them. From then on, under Parliament, the Protectorate, and later under the Restoration, he was regularly employed as code-breaker for the government. Despite requests from correspondents, Wallis never disclosed his techniques, but it is likely they were based on algebra, the cornerstone of his mathematical approach.

As a student of Holbeach and a graduate of Emmanuel College, Wallis was well integrated into Puritan circles and was a dyed-in-the-wool Parliamentarian from the earliest days of the conflict with Charles I. It

was not until 1644, however, that the young clergyman had his chance
to play a significant role in the great events of the day, when he was
called upon to give testimony at the trial of Archbishop William Laud.
As Archbishop of Canterbury, Laud had been the head of the Anglican
Church under Charles I, and the chief nemesis of the Puritans. In publicly testifying against him, Wallis showed himself a rising star in the
Presbyterian party, which led Parliament's struggle against the king in
the early years of the civil war. Confirmation of his status came that
same year, when he was appointed secretary to the Assembly of Divines
at Westminster. The assembly was the creation of the Long Parliament,
and was charged with coming up with a plan for one or more new churches
to replace the established Church of England. Like Parliament itself,
the assembly was dominated by Presbyterians, who wanted to abolish
bishops and replace them with a hierarchy of church elders ("presbyters"). Wallis, a bright, young, and well-connected minister with impeccable Presbyterian credentials, must have seemed an excellent choice to
run the assembly meetings.

Writing decades later, with both the king and the Anglican Church
safely restored, Wallis tried to downplay the radicalism of the Assembly
of Divines and minimize his role within it. But the facts speak for themselves: after years of intense debates, the Westminster Assembly recommended abolishing the episcopacy (i.e., bishops), the defining feature of
the Anglican Church and the emblem of its alliance with the king. The
diminishing clout of the Presbyterians and the ultimate restoration of
the Church of England ensured that the assembly's recommendations
were never put into practice. In their time, however, the proposals were
radical enough, representing a plan to make Church governance more
democratic and to remove the Church from royal control. As secretary
to the assembly, Wallis was a full partner to this agenda, and was known
as such.

Though the Presbyterians were the acknowledged leaders of the Parliamentary faction in the early years of the fight against the king, the
increasing radicalism of the revolution soon left them behind. Independents saw little difference between the offices of presbyter and bishop,
and wanted to abolish both. Even more ominously, Ranters, Quakers,
and Diggers threatened the very foundations of the social order. It was

enough to make the respectable Presbyterians wish for the bad old days of king and bishops, a time of law and order that now seemed infinitely preferable to the wholesale chaos they saw around them. And so, the Presbyterian radicals of the 1640s became the conservatives of the 1650s, fighting a rearguard action against the forces of subversion that they had unleashed years before. Wallis's own career traced a similar arc. In 1648 he signed a remonstrance to the army to protect the king's life, an honorable act that did nothing to save the king from execution. In 1649 he joined other London ministers in signing *A Serious and Faithful Representation*, protesting Pride's Purge, the army's expulsion from Parliament of members it considered too moderate. The purge was a crime, Wallis and his fellows declared, worse than any the king had perpetrated. Besides, they argued, with much revisionist hindsight, the Parliamentarians of 1640 had never dreamed of depriving the king of his rights and regal authority.

Brave gesture though it was, the *Representation* did nothing to save the king or the Presbyterian party. Once the heart and soul of the Parliamentary, the Presbyterians were a spent political force. Considered too conservative by the Independents and radicals who now dominated the commonwealth, they were also distrusted by the Royalists, who blamed them for making war on the king and unleashing the fury that was sweeping across the land. Unable to chart a path of their own, many Presbyterians quietly went over to the Royalist side, in the hope that their past misdeeds would be forgiven once the king was restored to the throne.

The decline in his party's fortunes left Wallis in an uncomfortable situation in London. He was still minister of St. Martin's Church in Ironmonger Lane, but generous patrons were harder to come by, now that the political climate had turned against the Presbyterians, and his ability to participate in the great events taking place around him was at an end. Even his personal safety could not be guaranteed at a time when Presbyterians were being accused of treason to the king by Royalists and treason to the cause by Independents and radicals. But it was at this juncture, when there seemed no future for him in London, that Wallis was granted an opportunity to leave it all behind and start afresh: on June 14, 1649, only months after signing the *Serious and Faithful*

Representation, he was appointed to the exalted position of Savilian Professor of Geometry at Oxford.

A MINISTER AND A PROFESSOR

To say that Wallis's appointment was a surprise is a vast understatement. Up until the previous year, the post of Savilian Professor had been held by Peter Turner, an accomplished and respected mathematician. But Turner, like many of his university colleagues, was a Royalist, and when in 1648 Parliament turned its attention to reforming the universities, he was ejected from his professorship, leaving the Savilian chair vacant. For his replacement, the authorities were looking for a scholar with strong Parliamentary credentials, and in that regard Wallis fit the bill. But Wallis was hardly qualified. Unlike Turner, who had been professor of geometry at Gresham College in London before coming to Oxford, Wallis had no track record in either teaching or publishing in mathematics. All he had were the accounting tricks he had learned from his younger brother in his youth, and years of haphazard readings in mathematics that he had pursued on his own and without guidance. The only mathematical work he had to his name was a treatise on angular sections, a topic far removed from the frontiers of the field, and that in any case remained unpublished. This is not the background one would expect of someone appointed to be professor of geometry at Oxford.

Wallis was appointed for political reasons, no more, no less, and it is safe to say that no one expected he would become a serious mathematician. How someone with so little to recommend him secured such a prize is a mystery to us today, but Wallis was nothing if not enterprising. Only a few years removed from his studies in Cambridge, he had managed to thrust himself to the center of national politics by becoming secretary to the Westminster Assembly of Divines. Now, when all avenues of advancement seemed to be closed to him, he bested the feat by winning the most desirable mathematical position in the land. It is possible, as his contemporary the antiquarian Anthony Woods hints, that the good relations he enjoyed with Oliver Cromwell played a role in the appointment. Cromwell knew and respected Martin Holbeach, his sons' teacher, and may have been impressed by the schoolmaster's high opinion of Wallis. Still, in 1649, Cromwell was just a general, not lord protector, and may not have had much influence in the matter. What is clear is that Wallis, thanks to a remarkable talent for getting by that shines brightly three and a half centuries later, had managed to extract himself from a difficult situation and land a brilliant prize. In London he was forever associated with the beleaguered Presbyterian faction, but in Oxford he was viewed simply as a Parliamentarian backed by the government that had installed him. And whereas in London he was viewed as a politician, and evaluated accordingly, at Oxford he would be judged as a scholar leading a life of contemplation removed from the tumult of revolutionary London. All he needed now was to become in reality what he already was in name: a mathematician.

This he accomplished with remarkable rapidity. Already in 1647 he had read William Oughtred's popular algebra textbook, *Clavis mathematicae* ("The Key of Mathematics"), and elaborated on it the following year by writing his own *Treatise on Angular Sections*, which was eventually published nearly forty years later. Now, as Savilian Professor, he recognized that "what had been before a pleasing Diversion was now to be my serious study," and embarked on a systematic self-education in the most current mathematical work. Wallis the dilettante, only minimally knowledgeable about modern mathematics, nonetheless absorbed the sophisticated mathematical works of Galileo and Torricelli, Descartes and Roberval. Within a few years he had not only mastered the work of his continental colleagues, but had embarked on his own program of mathematical research. In 1655 and 1656, six years after his appointment, he published two mathematical treatises of striking originality, one entitled *De sectionibus conicis* ("On Conic Sections"), the other *Arithmetica infinitorum* ("The Arithmetic of Infinites"). The works resonated far and wide in the European mathematical community, and were read from Italy to France to the Dutch Republic.

How did Wallis, the Puritan minister, manage to reinvent himself as an academic mathematician of international standing? His innate mathematical talent undoubtedly had much to do with it, as did his prodigious capacity for concentrated study and hard work. But there was more: Wallis had maneuvered deftly, relying on a vast web of connections and friends in high places. He had also shown himself flexible in his loyalties and ideological commitments: the doctrinaire Presbyterian of Emmanuel College and the Westminster Assembly of Divines was gone forever, replaced by a moderate clergyman content to profess loyalty to whoever was in power, whether Cromwell, the restored Stuart kings, or (after the Glorious Revolution of 1688) William and Mary. Indeed, when he wrote his autobiography toward the end of his life, he tried to paper over the rebellious politics of his youth by arguing, rather disingenuously, that the term *Presbyterian* referred to respectable clergymen who were opposed to the radical Independents, not to the Anglican bishops. His fluid allegiances and talent for backroom deals served Wallis well in the years following his appointment to the Savilian chair: in 1658 he was elected to the position of "keeper of the archives" (*custos archivorum*) of the University of Oxford, in a dubious process that elicited strong protest from his colleague at Oxford's Bodleian Library, Henry Stubbe, who was Hobbes's chief supporter at Oxford. In 1660 he was confirmed in his positions by the restored monarch, Charles II, and was later given the honorary title of "royal chaplain."

Wallis's flexibility of conscience was sure to elicit scorn from his famous rival Hobbes, who never wavered from his belief in a dictatorial Leviathan state, remaining obstinately true to his conclusions through thick and thin, with utter disregard for the mounting hostility of his peers. He was denounced as an atheist and a materialist, "a pander to bestiality" whose "doctrines have had so great a share in the debauchery of his generation that a good Christian can hardly hear his name without saying of his prayers." It was only his powerful patrons and the king's regard for his old tutor that saved Hobbes when Seth Ward, now the Bishop of Sarum, introduced a motion in Parliament that he be burned at the stake as a heretic. But shunned and abused as he was, Hobbes bore it all with fortitude, never deviating from his views. For the adaptable Wallis, and his talent for reinventing himself according to the prevailing political winds, he had nothing but contempt.

Inflexible and uncompromising, Hobbes's personal character mirrored his philosophical and mathematical views. In Euclidean geometry he recognized a system that, like him, was rigid and unyielding. Its critics were mere fools and knaves—and that was how Hobbes liked it. The opportunistic Wallis, meanwhile, had little interest in the grand claims of geometry as the embodiment of reason and a model for absolute truth. For him, mathematics was a pragmatic tool for obtaining useful results. He cared little if his proofs did not rise to the lofty level of certainty demanded by Euclidean mathematics. All he wanted were theorems that were sufficiently "true" for the business at hand. And if, in order to arrive at his results, he violated some of the cherished tenets of classical geometry, then those cherished tenets would simply have to give way. Traditional geometers might object to the notion that a plane is composed of an infinite number of lines, on the grounds that it violated well-known and ancient paradoxes. But if such an assumption proved effective in Wallis's calculations (as indeed it did), then he cared nothing for their objections. If some tweaking of principles was required in order to arrive at the desired end, then Wallis was happy to do it, in mathematics as in life.

For Wallis, Hobbes's rigidity was pedantic, intemperate, and ultimately self-defeating. He despised it in the man and he rejected it in his philosophy, but more than anything he considered it politically dangerous. A dogmatism that acknowledges only one single truth and denies the legitimacy, and even the possibility, of dissent, Wallis believed, would never bring about the civic peace that Hobbes sought. As Wallis saw it, inflexible dogmatism by the state would breed inflexible dogmatism and even fanaticism among its opponents, which in turn would lead to civil war and social and political chaos—precisely the outcome Hobbes was trying to guard against.

In fact, the foremost concern of both Wallis and Hobbes was the same: preventing a descent into the anarchy and chaos of the Interregnum. Both Wallis and Hobbes feared the Diggers of the world, and were equally intent on preserving the established order. They just differed sharply on the means to do so. Hobbes believed that the only way to preserve order was to establish a totalitarian state with absolutely no room for dissent. Wallis believed that the way forward was to allow for dissent within carefully prescribed limits, which would permit people to disagree and still preserve their common ground.

We do not need to make any inferences about Hobbes's political views or the role of mathematics within them, because he wrote it all down in beautiful prose. Wallis, in contrast, wrote extensively on

Iohannes Wallis, S.T.D.
Geometriæ Professor Savilianus Oxoniæ

John Wallis in 1670, at the height of his battle with Hobbes. Engraving by
William Faithorne. (Photograph courtesy of the National Portrait Gallery, London)

mathematics and authored many religious sermons over the years, but never claimed the mantle of philosopher. To piece together his views on political order, we need to move beyond his personal writing and toward the broader circle in which he moved. In his university days, and in the early days of struggle against the king, this meant the coterie of Presbyterian divines who dominated the Parliamentary party. But beginning in the mid-1640s, Wallis became a leading member of a different and far more diverse group. It met regularly in private homes in London and Oxford throughout the Interregnum and was known by different names at different times. Sometimes it was the "Invisible College"; at other times it was the "Philosophical Society." In 1662 the returning monarch, Charles II, finally gave it official recognition, a charter, and a name: the Royal Society of London.

SCIENCE FOR A GLOOMY SEASON

Three and a half centuries after its founding, the Royal Society is among the most august scientific institutions the world has ever known. To say that a list of its past fellows includes some of the greatest scientists in history is an understatement. If one counts the foreign fellows, it contains well nigh all of them. Robert Boyle (1627–91), of "Boyle's Law" fame, was one of the Society's founders and the most influential among the early fellows. Isaac Newton (1643–1727), often considered the first modern scientist, and whose *Principia mathematica* of 1687 revolutionized physics, astronomy, and even mathematics, was president of the Society from 1703 to his death in 1727. The Frenchman Antoine-Laurent Lavoisier (1743–94), founder of modern chemistry, was a foreign fellow, as was the American founding father Benjamin Franklin (1706–90). In later years there was Charles Babbage (1791–1871), designer of the first programmable computer; and William Thomson, Lord Kelvin, founder of the science of thermodynamics and Society president from 1890 to 1895. Charles Darwin (evolution), Ernest Rutherford (structure of the atom), Albert Einstein (relativity), James Watson (DNA), Francis Crick (also DNA), and Stephen Hawking (black holes) were or still are fellows. This is but a small selection of the most famous names among the fellows, but it is sufficient to get the picture: anybody who was anybody in the history of modern science was a fellow of the Royal Society.

But in 1645, when Wallis began attending informal meetings held by a group of gentlemen interested in natural philosophy, all this lay far in the future. The purpose of the meetings, as the Society's first historian, Thomas Sprat, wrote some years later, was not to found a scientific academy, and advancing the frontiers of knowledge was a secondary concern. "Their first purpose," Sprat reports, "was no more than breathing freer air, and of conversing in quiet with one another, without being engaged in the passions and madness of that dismal age." At a time when Royalists and Parliamentarians, Presbyterians and Independents, Puritans and Enthusiasts, property owners and tenants, were all at each other's throats, these men were seeking an escape. They found it in the study of nature.

"For such a candid and unpassionate company as that was," Sprat reflected, "and for such a gloomy season, what could have been a fitter subject to pitch upon than *Natural Philosophy*?" To discuss theological questions or "the distresses of their country" would have been too depressing. But nature could distract them, "draw their minds off past and present misfortunes," give them a sense of control in a world gone mad, and make them "conquerors over things." Their meetings were a space in which they could converse quietly, voice opposing views without shouting one another down, and find common ground despite disagreements. Amid the furor, fanaticism, and intolerance of revolutionary England, they were seeking a safe haven of tolerance to pursue a subject they believed would benefit all Englishmen, if not mankind. They called it "natural philosophy," and we call it science.

Wallis, by his own testimony, had already encountered the New Philosophy in his Cambridge days. Now, with his new companions, he began to pursue it systematically. Meeting weekly at the home of one of their members or at Gresham College, they discussed and experimented on the entire array of new ideas and discoveries that were shaking the foundations of the medieval order of knowledge. Wallis lists them all:

Physick, Anatomy, Geometry, Astronomy, Navigation, Staticks, Magneticks, Chymicks, Mechanicks . . . the Circulation of the Blood, the

Valves in the Veins, the Copernican Hypothesis, the Nature of Comets, and New Stars, the Satellites of Jupiter, the Oval Shape (as it then appeared) of Saturn, the spots in the Sun, and its Turning on its own Axis, the Inequalities and Selenography of the Moon, the several phases of Venus and Mercury, the improvement of Telescopes, and grinding of Glasses for that purpose, the Weight of Air, the Possibility or Impossibility of Vacuities, and Nature's Abhorrence thereof; the Torricellian Experiment in Quicksilver, the Descent of heavy Bodies, and the degrees of acceleration therein.

There were only two fields, Wallis explains, that were intentionally left out: "Theology and State affairs."

Wallis took part in the meetings in London for several years, even as he continued his career as a Presbyterian stalwart, protesting the king's execution and the army's purge of Parliament. It might be that, as he wrote years later, the apolitical experimentalists provided him with a welcome refuge from the dogmatic intolerance of Interregnum politics. It is just as possible that he was hedging his bets, hoping that his association with the natural philosophers would help him find a measure of security and success if the Presbyterians' power collapsed. That, in any case, is what happened. Wallis, who had been a mere dabbler in mathematics, started studying more advanced texts, which almost certainly played a part in his surprise appointment to the Savilian chair at Oxford.

The move to Oxford did not end Wallis's involvement with the group. Several other members ended up in Oxford around the same time, and together with some old Oxonians, they established the Oxford Philosophical Society and met regularly at the home of Robert Boyle. "Those in *London*, Wallis recalled, "continued to meet as before (and we with them, when we had occasion to be there;) and those of us in *Oxford* . . . continued such meetings in *Oxford*; and brought those Studies into fashion there." The two groups interacted closely, and when Charles II chartered the Londoners, the Oxford group was included, its members becoming founding fellows of the Royal Society. Wallis, a moving spirit behind the activities of both groups, became a prominent member of the new organization.

Under the king's patronage, the Royal Society became a trend-setting scientific organization and, along with the French Royal Academy of Sciences, a model for scientific institutions in Europe and beyond. Its regular meetings in these early years were devoted to public experiments in optics, the structure of matter, the reality of a vacuum, and telescopic observations, among other topics, executed by the Society's curator of experiments, Robert Hooke. Most famously, Robert Boyle's experiments with the air pump, in which he investigated the structure and composition of air, were conducted in public laboratories of the Royal Society in front of numerous witnesses. In 1665 the Society's secretary, Henry Oldenburg, launched *Philosophical Transactions of the Royal Society of London*, one of the first scientific journals, and certainly the longest-running in the world. *Philosophical Transactions* reported not only the investigations of the Society's fellows but also studies conducted by others, making the Society a world center of scientific research.

Some of the practices of the early Royal Society might seem peculiar to a modern scientist. For example, there was little distinction between what today would be considered amateurs and professionals, and the pages of early *Philosophical Transactions* were filled with reports of unusual weather phenomena and the births of monstrous or malformed livestock. Social rank also mattered a great deal in the Society, and quite a few prominent gentlemen owed their fellowship to their illustrious lineage rather than to any scientific distinction. Also puzzling from our perspective is that the experiments were performed in public, that is, in front of an audience of Society fellows and sometimes other notable guests. All those present would then discuss what they had seen, examining its meaning and significance. To a modern scientist this would seem more like a circus performance than a proper scientific experiment.

Some of the differences between the early Royal Society and modern scientific procedures can be attributed to the fact that science in the seventeenth century was still young, and its practices still very much in flux. The professional career scientist is a creation of the nineteenth, not the seventeenth, century. Other differences are due to the Royal Society's viewing itself as much more than a scientific institution of the kind we know today. A modern scientific institute or university department is concerned exclusively with scientific research and education, its success measured by the number and quality of its publications and innovations. The Royal Society also emphasized its research and innovation, always insisting on the usefulness of its discoveries. But in addition to this, it took upon itself a mission unlike that of any of its modern counterparts: to provide a model for the operation of the state as a whole.

This mission has its roots in the meetings of the London group, back in the 1640s. Outside their meeting hall the group's members might be radicals or moderates, Presbyterians or Independents, Parliamentarians or even Royalists, all engaged in a life-and-death struggle for domination. But inside their meetings, none of that mattered: irrespective of their religious and political affiliations, they could all pursue the investigations of nature in peace and civility. "It was *Nature* alone," wrote Sprat of those early meetings, that ". . . draws our minds off from past, or present misfortunes . . . that never separates us into mortal Factions; that gives us room to differ, without animosity; and permits us to raise contrary imaginations upon it, without any danger of *Civil War.*" In pursuing nature, Wallis, Boyle, and their associates created a safe space where even disagreements could be managed in peace and civility. It was a welcome relief from the cutthroat politics of the Interregnum.

But what started as a simple refuge ultimately developed into an ideal: if reasonable men of different backgrounds and convictions could meet to discuss the workings of nature, why could they not do the same in matters that concerned the state? Why could Parliamentarians and Royalists not resolve their differences in peace and civility rather than murdering each other on the battlefields of northern England? Why could Independents, Presbyterians, and Anglicans not come to a reasonable agreement on Church government instead of each trying to impose their own system and suppressing all others? The harmony that prevailed in the natural philosophers' meetings, even among men who disagreed sharply, seemed to hold an important lesson for the entire English body politic. For as Sprat put it, in those meetings, "we behold an unusual sight to the *English Nation*, that men of disagreeing parties, and ways of life, have forgotten to hate, and have met in the unanimous advancement of the same *Works* . . . For here they do not only endure each other's presence without violence and fear; but they *work* and *think* in company, and confer their help to each other's *Inventions."*

In the harsh climate of Interregnum politics, Wallis and his fellows reveled in their ability to conduct their business in peace, cooperate despite disagreements, and together advance the cause they all cherished. By the time they emerged from the shadows and were officially chartered by Charles II, they were ready to spread the word, and use their experience to reconstitute the entire body politic. The dogmatism of the preceding decades would be replaced by the moderation and open-mindedness that characterized their meetings and their science. The hubris of the fanatics would be replaced by the modesty of the experimenter, passions by rational debate, and the intolerance of sects by the tolerance of different but reasonable men working together for a shared cause.

Presenting itself as a model for the state, the Royal Society tried to be as inclusive as possible. There was nothing democratic about it, to be sure, and members of the lower classes were unwelcome in its halls just as they were unwelcome in the political class. Wallis, Boyle, and their fellows feared and distrusted the common people, convinced that the only way to achieve peace and order was to restore the authority of the propertied classes. But when it came to gentlemen, the Society sought to set an example of openness, and this meant accepting into their ranks men of even modest accomplishment. For Society fellows to have presented themselves as "professionals," to the exclusion of "amateurs," would have smacked too much of the sectarianism of the past, in which one group sets itself up as the judge of all others.

The public experiments of the early days also played a role in the Royal Society's political mission, serving as an example of how reasonable men of good faith could discuss difficult issues and come to an agreement. The model was to be the private meetings of the Society's founders during the Interregnum, in which they experimented and debated, offering different interpretations of what they were observing. In the end, though, they would arrive at some interpretation they could agree on—even if it left many questions unanswered. But for such discussions to take place, now that the Society was an official institution, it would not do to conduct the experiments in the privacy of a secluded laboratory. If members were to form an opinion, they would have to observe the proceedings themselves. It was therefore essential that the experiments be conducted before witnesses of unimpeachable character, most often other fellows, who could then discuss what they had seen and come to an agreement on what had occurred. A modern-day laboratory, in contrast, does not carry the ideological burden of the early Royal Society. It relies exclusively on the testimony of experts, safely assuming that laymen would not comprehend the proceedings anyway.

Not all forms of natural philosophy were equally suited for the Society's goals of promoting peace, tolerance, and public order. Particularly suspect were grand philosophical systems that claimed to arrive at indisputable truths through the power of pure reason. One such system, which was very much on the minds of the Society's founders, was Cartesian philosophy (named after its originator, René Descartes), which was sweeping the Continent at that very time. In his writings, Descartes purported to dismantle all unsubstantiated presuppositions, reducing all knowledge to a single unshakeable truth: "I think, therefore I am." From this rock of certainty he then recreated the world through rigorous step-by-step reasoning, accepting the validity of only clear and distinct ideas. And since his reasoning was flawless, Descartes (and his followers) argued, his conclusions must inevitably be true.

Boyle, Wallis, Oldenburg, and the other leaders of the early Society were deeply impressed by Descartes, but also very critical of his approach and conclusions. They were even more concerned with another system anchored in pure reasoning that was lurking in their own backyard, and that of course was Hobbes's philosophy. Hobbes and Descartes differed radically on many critical issues, but this much they had in common: both believed that their system was structured like Euclidean geometry, founded on self-evident assumptions and proceeding through rigorous reasoning to truths. And it was precisely this unquestioning confidence in the validity of their systematic reasoning and the absolute truth of their conclusions that the founders of the Royal Society found particularly dangerous.

The problem with dogmatic philosophy, Sprat explained in his *History of the Royal Society*, "is that it commonly inclines such men, who think themselves already resolv'd, and immovable in their opinions, to be more imperious, and impatient of contradiction." Such an attitude is detrimental to science because "it makes them prone to undervalue other men's labours, and to neglect the real advantage that may be gotten by their assistance. Least they should seem to darken their own glory." It "is a Temper of mind, of all others the most pernicious," Sprat continued, and one to which he attributes the "slowness of the increase of knowledge amongst men." Even worse, this kind of arrogance easily leads to the subversion of the state: "The *reason* of men's contemning all *Jurisdiction* and *Power* proceeds from their idolizing of their own *Wit* . . . they suppose themselves *infallible*." This leads inevitably to sedition, because "the most fruitful parent of *Sedition* is *Pride*, and a lofty conceit of men's own *wisdom*; whereby they presently imagine themselves sufficient to direct and censure all the *Actions* of their *Governors*."

Sprat was only twenty-eight when he was elected fellow in 1663, a young and not particularly distinguished man who was probably recruited for the express purpose of writing the Royal Society's history. But if Sprat was a relative nonentity at the time, the men who commissioned him to write were the Society's greatest men. These included the Society's president, Lord Brouncker; its secretary, Henry Oldenburg; and its leading scientist, Robert Boyle, all of whom reviewed and corrected Sprat's text to make sure it accurately presented their views. As a result, *History of the Royal Society* is not just a summary of Sprat's private reflections, but a public statement of the goals and purpose of the Royal Society as understood by its leaders at the time. And when it came to their views on dogmatic philosophies, their verdict was clear: dogmatism leads to sedition and subversion of the state, and was *not* the kind of approach that would be practiced in the Royal Society.

The alternative to the dogmatic rationalism of Descartes and Hobbes, the founders of the Royal Society believed, was experimental philosophy. Instead of pride, experimentalism bred humility, and whereas the rationalist philosophies led to pettiness and envy of rival philosophers, experimentalism fostered cooperation and mutual trust. Most important, instead of sedition and subversion, "the influence of experiments is *Obedience to the Civil Government*." Unlike the rationalist philosopher, the experimentalist never claims he has discovered the only true system or that his results are absolutely and irrefutably true. Instead, making no assumptions about what he will find, he humbly proceeds from experiment to experiment, trying to make sense of what he finds. His conclusions are always the best that he can supply at the moment, but can always be overturned by the next experiment. Not for him are Hobbes's bold pronouncements about matter, human nature, and the only viable commonwealth. To the contrary, he proceeds slowly, conducting many different experiments many times over, and only then will he venture, carefully and somewhat reluctantly, to provide a provisional interpretation of the results.

Experimentalism is a humbling pursuit, very different from the brilliance and dash of systematic philosophers such as Descartes and Hobbes. It is, wrote Sprat, "a laborious philosophy . . . that teaches men *humility* and acquaints them with their own *errors*." And that is precisely what the founders of the Royal Society liked about it. Experimentalism, as Sprat noted, "removes all haughtiness of mind and swelling imaginations," teaching men to work hard, to acknowledge their own failures, and to recognize the contributions of others. This is precisely the attitude the founders of the Royal Society hoped to engender in the body politic as a whole. In place of the intolerant fanaticism of the parties and sects that had thrown the commonwealth into violence and chaos, experimentalism would breed moderation, cooperation, respect for differing opinions, and ultimately civic peace.

When members of the Royal Society celebrated the glories of the experimental method, they also celebrated the man whom they considered the founder of it all, the "one great man who had the true imagination of the whole extent of this Enterprize." He was Francis Bacon, Lord Chancellor to James I, who in his retirement had authored some of the most influential works ever on proper scientific method. In contrast to his younger contemporary Descartes, who had argued that true knowledge must be based on clear and rigorous reasoning, Bacon had insisted that true knowledge of nature could be acquired only by observation, experimentation, and the careful gathering of facts. To the Royal Society, Bacon was the prophet of the experimental method, and the spiritual father of the Society itself, though he died many years before its founding. In fact, the Society considered itself the true incarnation of Bacon's "Salomon's House," a state institution for the study of nature that he proposed in his utopian work *New Atlantis*.

There is irony here, because Bacon's secretary in his final years was none other than Thomas Hobbes. As an avowed rationalist, Hobbes had derided the value of experiments in his dispute with Robert Boyle, and his thinking was not much influenced by his distinguished employer (except perhaps in his abiding interest in the natural sciences). But there was no getting around the fact that for all their idolization of Bacon, none of the Society grandees had actually known the Lord Chancellor, whereas their enemy Hobbes had been his intimate companion.

The brilliance of Bacon's reputation has hardly diminished, even to this day. Though not a creative scientist himself, he is nevertheless considered one of the crucial figures in the scientific revolution, whose writings made possible the growth and expansion of science. Bacon provided a brilliant defense of the experimental method, which had been viewed as suspect during the centuries in which Scholastic dispute and reliance on ancient authority were considered the proper path to true knowledge. He provided a road map for the development of experimental science, advocating for the systematic collection of data by a multitude of field-workers, and its concentration in a centralized institution for systematic evaluation. More than anything, perhaps, he made the experimental method respectable.

Long before Bacon's time, there were always those who tried to extract the secrets of nature through the rough method of trial and error. Sometimes they succeeded brilliantly, as in the inventions of gunpowder and the compass, other times less so, as in the case of the alchemists, who built sophisticated laboratories equipped with chemicals and furnaces in their search for the elusive philosopher's stone. But any knowledge gained by these methods, even when it proved useful, was not considered appropriate for teaching in institutions of higher learning. It was "rude" and "mechanic," associated with the lower classes, who dirtied their hands and worked for a living. No self-respecting gentleman would ever stoop to engage in such work, for fear of being tainted by its plebeian association. True knowledge, worthy of academic study, was to be found in the writings of the great masters of the past, or derived from them through exacting logical reasoning. Experimental results were not considered knowledge at all, since they relied on the notoriously unreliable senses and did not therefore rise to the required level of certainty. Bacon, almost single-handedly, demolished this perception. Here was no less a personage than the Lord Chancellor of England promoting experi-

mentalism as the proper path to true knowledge. At a stroke, the uncouth practices of "rude mechanics" became a worthy pursuit for the
intellectually curious gentleman.

There is one aspect of Bacon's methodology, however, that has often
been criticized: his belittlement of mathematics as a tool of science. It is
not that he ignored mathematics completely, since he did acknowledge
that objects in the world had quantity, and mathematics was the science
of quantity. But Bacon thought that mathematical knowledge was far too
general to be of serious use. "It is the nature of the human mind," he
wrote, "to delight in the open plains (as it were) of generalities, rather
than the woods and inclosures of particulars," and mathematics was the
best field to "satisfy that appetite." Such an approach, however, is "to
the extreme prejudice of knowledge," because all knowledge worth pursuing lies in the particulars of the tangled woods, not in the generalities
of the open plains. Mathematics can be useful, Bacon concedes, but
only as the handmaiden of the experimental fields, not as a science unto
itself. Nothing could be worse for the growth of knowledge than "the
daintiness and pride of mathematicians, who will need have this science
almost domineer over Physic."

Bacon's suspicion of mathematics as a tool for comprehending the
world is not hard to understand. For mathematics to describe nature
correctly, nature must be mathematical—that is, structured according
to strict mathematical principles. If that is the case, then all one needs
in order to gain insight into the workings of nature is to follow the rules
of rigorous mathematics, and all observations and experiments are su-
perfluous. But Bacon made no such assumption. There is no way of
knowing how the world is structured, he believed, until one engages in
careful and systematic observations. The idea that one can deduce the
workings of nature by mere mathematical reasoning is a dangerous illu-
sion based on unwarranted pride, and is bound to lead any scientist
astray.

Bacon's warning against the "daintiness and pride" of mathematicians was not lost on his followers, the founders of the Royal Society.
Although the Society was officially described as a "Colledge for the
Promoting of Physico-Mathematicall Experimentall Learning," in practice the "mathematicall" studies were strictly subordinated to the

"experimentall" ones. For the Society leaders shared Bacon's concern that mathematics breeds pride and makes it easy to assume that God created the world according to rigid mathematical strictures. Like Bacon, they were worried that mathematical reasoning would lure scientists away from the laborious work of experimentation.

But the Royal Society founders had other concerns, concerns that went beyond Bacon's warning half a century before. Mathematics, they believed, was the ally and the tool of the dogmatic philosopher. It was the model for the elaborate systems of the rationalists, and the pride of the mathematicians was the foundation of the pride of Descartes and Hobbes. And just as the dogmatism of those rationalists would lead to intolerance, confrontation, and even civil war, so it was with mathematics. Mathematical results, after all, left no room for competing opinions, discussions, or compromise of the kind cherished by the Royal Society. Mathematical results were produced in private, not in a public demonstration, by a tiny priesthood of professionals who spoke their own language, used their own methods, and accepted no input from laymen. Once introduced, mathematical results imposed themselves with tyrannical power, demanding perfect assent and no opposition. This, of course, was precisely what Hobbes so admired about mathematics, but it was also what Boyle and his fellows feared: mathematics, by its very nature, they believed, leads to claims of absolute truth, dogmatism, threats of tyranny, and, all too easily, civil war.

Yet, despite the ideological and political dangers, mathematics could not be simply dispensed with. Some of the greatest accomplishments of the New Philosophy were heavily mathematical. Advancements in medicine, such as Harvey's discovery of the circulation of the blood, were certainly experimental, as were the barometric measurements of atmospheric pressure known as the Torricellian experiment, and William Gilbert's investigations into the nature of magnetism. But the greatest scientific triumphs of the age were indeed in astronomy, and these were deeply indebted to mathematics.

What, then, could the Royal Society leaders do? They could not simply ignore the brilliant contributions that mathematics had already made to science, or the strong indications that the former would continue to play a central role in the latter's advancement. But how could the Society embrace the important contributions of mathematical science and yet avoid its dangerous methodological, philosophical, and political implications? It was a conundrum that left the Society with an ambivalence toward mathematics that characterized its science for many years. And no one felt this conflict more keenly than John Wallis.

9

Mathematics for a New World

AN INFINITY OF LINES

Wallis was the only mathematician among the Society's founders, and it therefore fell to him to address the problematic status of mathematics. He fully shared his associates' abhorrence of dogmatism, and in his autobiography he took pride in his moderation and openness to diverse opinions, even when they conflicted with his own views. "It hath been my endeavour all along," he wrote by way of a summary, "to act by moderate principles, between the extremities on either hand . . . without the fierce and violent animosities usual in such cases against all that did not act just as I did, knowing that there were many worthy persons engaged on either side."

Yet, as a mathematician and Savilian Professor, Wallis was committed to a field that had traditionally prided itself on its inflexible methodology and the absolute and incontestable truth of its results. How was one to reconcile this with the moderation and flexibility he cherished as a member of the Royal Society? Wallis's solution was simple, and radical: he created a new kind of mathematics. Unlike traditional mathematics, this new approach would proceed not through rigorous deductive proof but through trial and error; its results would be extremely probable but not irrefutably certain, and they would be validated not through "pure reason" but by consensus, just like the public experiments conducted at the Royal Society. Ultimately, his mathematics would be judged not by its logical perfection, but by its effectiveness at producing new results.

His mathematics, in other words, was not modeled on Euclidean geometry, the grand logical edifice that had inspired Clavius, Hobbes, and innumerable others over two millennia; it was, rather, designed to emulate the experimental approach practiced at the Royal Society. If Wallis succeeded, he would free mathematics of its association with dogmatism and intolerance and resolve the long-standing objections his fellows at the Society had toward the field. It would be a new "experimental mathematics," powerful and effective in the service of science, but serving as a model for tolerance and moderation rather than dogmatic rigidity. And at its very core would be the concept of the infinitely small.

The singular nature of Wallis's approach is apparent from the very first theorem of the very first work he wrote and published as Savilian Professor, "On Conic Sections" (*De sectionibus conicis*).

> I suppose, to begin with (according to Bonaventura Cavalieri's *Geometry of Indivisibles*) that any plane is made up, so to speak, of infinite parallel lines. Or rather (as I prefer) of an infinite number of parallelograms of equal height, the altitude of each one being $\frac{1}{\infty}$ of the entire height, or an infinitely small aliquot part (the sign ∞ denoting an infinite number); so that the altitude of all equal the height of the figure.

Immediately we are in the highly unorthodox world of Wallis's infinitesimal mathematics. Like Cavalieri and Torricelli before him, Wallis considered planes as quasi-material objects made up of an infinite number of lines stacked on top of one another, not as the abstract concepts of Euclidean geometry. That this conflicted with the classical paradoxes of Zeno, and with the problem of incommensurability, was obvious to any mathematician reading the tract, and both Hobbes and the French mathematician Pierre de Fermat were quick to point this out. But Wallis was unimpressed with these obvious criticisms. His notion that plane figures are composed of lines was derived from Cavalieri's famous analogy of a plane to a piece of cloth made of threads, as well as from the Jesuat's practice of viewing a plane as an aggregate of lines. He therefore simply referred the reader to Cavalieri, who supposedly had already dealt with all objections and moved on. Wallis even invented a sign to mark the number of infinitesimals that make up the plane and their magnitudes, respectively, ∞ and $\frac{1}{\infty}$.

With these basic tools in hand, Wallis then proceeds to demonstrate the power of his approach by proving an actual theorem:

> Since a triangle consists of an infinite number of arithmetically pro-
> portionate lines or parallelograms, beginning with a point and con-
> tinuing to the base (as is clear from the discussion): then the area of
> the triangle is equal to the base times half the altitude.

Needless to say, Wallis did not need to provide a complex proof in order to determine that the area of a triangle is half its base times its height. The purpose of the proof was not to prove the result, but quite the op-posite: to demonstrate the validity of his unconventional approach, by showing that it led to a correct and familiar result. Once he had estab-lished the reliability of his method, he could then use it to resolve more challenging and unfamiliar problems.

The statement that the lines composing a triangle are "arithmetically proportionate" calls for some explanation. What Wallis means is that if lines are drawn through a triangle parallel to its base, and if those lines are equally spaced along the triangle's height, then the lengths of the lines form an arithmetic progression. For example, if a line is drawn halfway between the apex of the triangle and the base, its length will be half that of the base, forming the arithmetical series $(0, \frac{1}{2}, 1)$, for the apex, the line, and the base, respectively. If the height is divided into three, and lines are drawn at the one-third and two-third marks, their lengths will form the series $(0, \frac{1}{3}, \frac{2}{3}, 1)$; if the height is divided into ten, the lengths will be $(0, \frac{1}{10}, \frac{2}{10}, \frac{3}{10} \ldots \frac{9}{10}, 1)$, and so on. This holds true regard-less of how many parts the height is divided into, as long as those parts are at an equal distance from one another. In his proof, Wallis assumes that this principle holds true even if the height is divided into an infinite number of parts.

He continues: "It is a well-known rule among mathematicians that the sum of an arithmetical progression, or the aggregate of all the terms, equals the sum of its extremes multiplied by half the total number of terms." This is a simple rule, familiar today to many a secondary school student. The sum of all the numbers from 1 to 10, for example, is 11 (that is, $1 + 10$) times 5 (half the number of terms in the sequence), that is, 55. Designating the infinitesimal magnitude of a single point by the letter "o," Wallis then uses the rule to sum up all the indivisible lines that compose the triangle:

> Therefore, if we consider the smallest term "o" (since we suppose that a point equals "o" in magnitude as well as zero in number), the sum of the two ends is the same as the largest term. I substitute the altitude of the figure for the number of terms in the progression, for this reason, that if we suppose the number of terms to be ∞, then the sum of their lengths is $\frac{\infty}{2}$ × *Base* (since the base is equal to the sum of the two ends).

Wallis is after the total length of all the lines that make up the triangle. Since they are infinite in number, and range from zero (or "o") to the length of the base, their combined length is $\frac{\infty}{2}$ × *Base*. He now multiplies this by the thickness of each line:

> But we suppose the thickness or altitude of each (line or parallelogram) to be $\frac{1}{\infty}$ × *the Altitude of the Triangle*; by which the sum of the lengths is to be multiplied. Therefore $\frac{1}{\infty}$ × A multiplied by $\frac{\infty}{2}$ × *Base* will give the area of the triangle. That is $\frac{1}{\infty} A \times \frac{\infty}{2} B = \frac{1}{2} AB$.

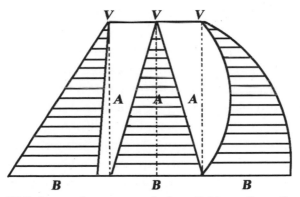

Figure 9.1. Wallis's triangles, composed of parallel lines. From *De sectionibus conicis*, prop. 3. (Oxford, Leon Lichfield, 1655)

And that is how Wallis calculated the area of the triangle: He summed up the lengths of all the component lines as an arithmetical progression, and then multiplied the sum by the "thickness" of each line. Arriving in this way at an equation that had ∞ in the numerator and ∞ in the denominator, he canceled them against each other and ended up with the familiar formula. *QED*.

Now, it is probably an understatement to say that no modern mathematician would follow Wallis in these wild and woolly calculations. Nor would many of his contemporaries, including all the Jesuits and Fermat, among others. Apart from the problematic assumption that a surface is made up of lines with a certain (very small) thickness, Wallis is also assuming without proof that the rules for summing up a finite series also apply to an infinite one. And if these unsubstantiated assumptions are not questionable enough, Wallis then casually divides infinite by infinity, or, to use his own notation, ∞ by ∞. In modern mathematics $\frac{\infty}{\infty}$ is undefined, for the simple reason that if $\frac{\infty}{\infty} = a$, then $\infty = a \times \infty$, and since any number multiplied by ∞ equals ∞, a can be any number. But Wallis treats $\frac{\infty}{\infty}$ as an ordinary algebraic expression, and cancels ∞ by ∞. When criticized by Fermat and others, Wallis seemed unconcerned with the logical difficulties of his procedures, refusing to concede any of their points. His approach, after all, was not meant to demonstrate his adherence to strict formal rigor. It was, rather, designed to make mathematics acceptable to his fellows at the Royal Society.

What did Wallis accomplish with his unconventional approach? For one, he posited that geometrical objects were objects "out there" in the world, and could be investigated as such, just like any natural object. This is the exact opposite of the traditional view that held that all geometrical objects should be constructed from first principles. It also runs counter to Hobbes's view that geometrical objects are perfectly known because we construct them. For Wallis, in contrast, the triangle already exists in the world, and the geometer's job is to decipher its hidden characteristics—just like a scientist trying to understand a geological rock formation or the biological systems of an organism. Drawing on common sense and intuition about the physical world, Wallis concluded that a triangle is composed of parallel lines next to each other, just as a rock formation is made of geological strata, a piece of wood is made of fibers, or (following Cavalieri) a piece of cloth is made of threads.

Since, according to Wallis, geometrical objects already exist in the world, it follows that mathematical rigor is completely unnecessary. In traditional geometry, in which one constructs geometrical objects from first principles and proves theorems about the relationships between them, logical rigor is indispensable. It is only a strict insistence on correct logical inferences, after all, that guarantees that the results are correct. The case, however, is very different if one is examining an object in the world, because it is external reality that decides whether a result is correct. An overinsistence on strict logical reasoning can be more a hindrance than a help.

Consider, for example, a geologist investigating a rock formation. He will certainly not throw his results out the window just because someone points out that his grant proposal has a spelling mistake or one of the measurements contains a tiny error. Rather, if the results correctly describe the rock formation—its structure, age, the manner it was formed, etc.—then the geologist will rightly conclude that the overall methodology must be correct as well, regardless of minute inconsistencies. The same is true for Wallis, who studied triangles as external objects no different, fundamentally, from a rock formation. It is all very well to insist on strict rigor, we can imagine Wallis thinking, but not if it gets in the way of our making new discoveries. Some mathematicians could and did gripe about his infinitesimal lines and his division of infinity by infinity, but to Wallis this was mere pedantry. He did, after all, arrive at a correct result.

Such casual disregard for logical rigor is a strange position for a mathematician to take, but Wallis had signaled his unusual outlook in *Truth Tried* in 1643. Rejecting the pure reasoning of Euclidean geometry, Wallis instead posited the triangle as an almost material object, one that could be intuited through the senses. For Wallis, the triangle could certainly be seen, its internal structure "felt," and if it could not quite be "tasted," then one got the feeling that it almost could be. "Mathematical entities exist," he wrote confidently in his *Mathesis universalis* of 1657, "not in the imagination but in reality."

Much had happened in Wallis's life in the years that passed between the publication of *Truth Tried* and his mathematical publications of the following decade. He had left his Presbyterian roots largely behind, moved from London to Oxford, and become a professional mathematician and Savilian Professor. But when it came to the question of how to acquire true knowledge, Wallis the distinguished mathematics professor was no different from Wallis the young Parliamentarian firebrand: the path lay not through abstract reasoning, but through material intuition, which "it seems not in the power of the Will to reject."

EXPERIMENTAL MATHEMATICS

Wallis's approach in *De sectionibus conicis* established geometrical objects as real bodies in the world, but it left open the question of how they should be investigated. In the proof of the area of the triangle, Wallis relied on material intuition to break down the plane into an infinity of parallel lines, and then sum them up. This proves to be effective for the problem at hand, but it is not a "method" applicable to a wide array of mathematical problems. In *Truth Tried*, Wallis suggested that a broader approach should rely on experiments, but it is far from clear what that means. How should one apply the experimental method, which relies on material scientific instruments and actual physical observations to abstract mathematical bodies such as triangles, circles, and cones? Wallis had an answer, and he gave it in the *Arithmetica infinitorum*, which was published alongside the *De sectionibus conicis* in 1656. It is widely considered to be his masterpiece.

"The simplest method of investigation, in this and various problems that follow, is to exhibit the thing to a certain extent, and to observe the ratios produced and compare them to each other; so that at length a general proposition may become known by induction." So wrote Wallis in proposition 1 on the first page of the *Arithmetica infinitorum*. The critical word here is also the last, *induction*, and the approach became known both to Wallis and to his critics as his method of induction. Today, mathematical induction is the name given to a perfectly rigorous and widely used method of proof that is taught to secondary school and university students. It consists of demonstrating a theorem for a particular case, say, n, and then proving that if it is true for n, it is also true for the case $n + 1$ (or $n - 1$), and consequently for all n's. This, however, was developed much later, and is not at all what Wallis had in mind. Because in the seventeenth century, and especially in seventeenth-century England, "induction" was associated with a particular scientific approach and a particular individual: Francis Bacon, Lord Chancellor to James I and prophet and chief advocate of the experimental method.

Bacon developed his theory of induction in his *Novum organum* (*The New Organon*) of 1620, his most systematic work on scientific method. He saw induction as an alternative to deduction, which, according to Aristotle and his followers in European universities, is the strongest form of logical reasoning. Deduction is the type of reasoning employed in Euclidean geometry and also in Aristotelian physics. It moves from the general ("all men are mortal") to the particular ("Socrates is mortal") and from causes ("heavy bodies belong in the center of the cosmos") to effects ("heavy bodies fall to the ground"). But Bacon argued that this kind of reasoning would never lead to new knowledge, because it made no room for the acquisition of new facts through observation and experimentation. Induction, for Bacon, was an alternative form of reasoning that, unlike deduction, could make use of experiments.

Induction was far from a new idea in the early 1600s. It was certainly known to Aristotle and other ancient philosophers, who considered it an inferior form of reasoning compared to deduction. Instead of moving from the general to the particular, induction does the reverse: it requires the gathering of many particulars and drawing from them an overarching rule. It follows that instead of proceeding deductively from causes to effects, induction starts with effects taken from the world around us and then, from them, infers the causes.

The pitfalls of inductive reasoning become clear if we consider the case of the black swan, a favorite cautionary tale of many philosophers. For many centuries Europeans had lived with swans and observed them, and all the swans they had seen were white. Using induction, they reasonably concluded that all swans were white. But when Europeans arrived in Australia in the 1700s, they made an unexpected discovery: black swans. It turns out that despite the innumerable particular observations by Europeans over many centuries, and despite the fact that every single observation was of a white swan, the rule that "all swans are white" was nevertheless false.

Writing in the early 1600s, Bacon knew nothing of black swans, but he was fully aware of the inherent uncertainty of induction. Yet he was undeterred. Aristotelian physics, he believed, was a well-constructed and elegant trap, logically consistent but completely divorced from the world. The only way to expand human knowledge of the world, he argued, was through direct engagement with nature, and that meant systematic observations and experimentation. Since these methods, he conceded, work by induction, they are vulnerable to the weaknesses of induction, and their conclusions are never absolutely certain. But if applied carefully and systematically, Bacon argued, with full awareness of its potential weaknesses, induction can ultimately lead to the advancement of human knowledge. It is the only way, according to Bacon, to study nature and uncover its secrets.

So when Wallis at the beginning of the *Arithmetica infinitorum* writes that he will proceed through induction, he is associating himself with a very particular philosophical enterprise: the experimental philosophy advocated by the late Sir Francis Bacon and later adopted and promoted by the founders of the Royal Society. Wallis had already demonstrated in *De sectionibus conicis* that he viewed mathematical objects as existing in the world, just like physical objects. In *Arithmetica infinitorum* he indicates how he is going to study them: through experiments. He will, in other words, study triangles, circles, and squares using the same method that his friend Robert Boyle used to study the structure of the air and his colleague Robert Hooke to study minute creatures under a microscope. When attempting to establish a mathematical truth, he will begin by trying it out on several particular cases and carefully observing the results of these "experiments." At length, after he has done this repeatedly for different cases, "a general proposition may become known by induction." Wallis had found the answer to his colleagues' suspicion of the mathematical method: he had developed an experimental mathematics to fit the spirit of the Society's experimental ethos. Rather than deduce universal laws that compelled assent and ruled out dissent, Wallis's mathematics would gather its evidence gradually, case by case, and slowly, cautiously arrive at general, and provisional, conclusions. For such is the way of the experimenter.

Wallis's experimental mathematics is the basic tool of the *Arithmetica infinitorum*, the foundation of his mathematical reputation. The subject of the work is very similar to Hobbes's most ambitious mathe-matical venture: determining the area of the circle. There is, however, a crucial difference between their projects. Hobbes tries actually to con-struct a square equal in area to a circle, using only the traditional Eu-clidian tools of straightedge and compass. He was doomed to fail, because the side of a square with the area of a circle with radius r is $\sqrt{\pi}r$, and π (as it was shown two centuries later) is a transcendental number, which cannot be constructed in this manner. Wallis, of course, does not try to construct anything. He instead tries to arrive at a number that will give the correct ratio between a circle and a square with a side equal to its radius r. Since the area of the square is r^2 and the area of the circle is πr^2, the number is π. Since π is transcendental, it cannot be described as a regular fraction or a finite decimal fraction. Nevertheless, at the end of the work, Wallis manages to produce an infinite series that allows him to approximate π as closely as he wishes:

$$\frac{4}{\pi} = \frac{3 \times 3 \times 5 \times 5 \times 7 \times 7 \times 9 \times 9 \times 11 \times 11 \times \ldots}{2 \times 4 \times 4 \times 6 \times 6 \times 8 \times 8 \times 10 \times 10 \times 12 \times \ldots}$$

Wallis begins his calculation of the area of a circle much as he began his calculation of the area of a triangle: Looking at one quadrant of the circle with radius R, he parses it into parallel lines as seen in Figure 9.2. The longest of these is R, and the others gradually become shorter and shorter until they reach 0 at the circle's circumference. Let us mark the longest line r_0, and the others r_1, r_2, r_3, and so on. Meanwhile, the area of the square enclosing the quadrant is also composed of an infinity of lines, but they are all the same length. Consequently, the ratio between the quadrant and the square is

$$\frac{r_0 + r_1 + r_2 + r_3 + \ldots + r_n}{R + R + R + R + \ldots + R}$$

The more lines we have in the quadrant and in the square—or, as we would say today, as n approaches infinity—the closer this number draws to the ratio between the area of the quadrant and the area of the enclos-ing square.

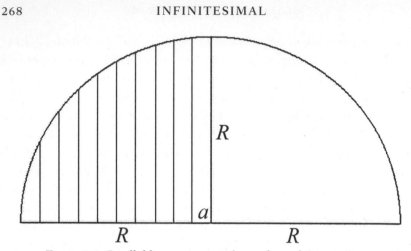

Figure 9.2. Parallel lines compose the surface of the quadrant.
After John Wallis, *Arithmetica infinitorum* (Oxford: Leon
Lichfield, 1656), p. 108, prop. 135.

Now, the precise length of each of the parallel lines *r* that compose
the quadrant is dependent on its distance from the first and longest line,
R. If we divide the distance *R* into *n* equal parts, and consider each equal
part a unit, then the length of the line closest to *R* is $\sqrt{R^2 - 1^2}$; the line
next to that will be $\sqrt{R^2 - 2^2}$, the one after that $\sqrt{R^2 - 3^2}$, and so on, until
we reach the circumference, where the last line is $\sqrt{R^2 - R^2}$, which is
zero. The ratio between the lines dividing the quadrant and the same
number of lines dividing the enclosing square will therefore be

$$\frac{\sqrt{R^2 - 0^2} + \sqrt{R^2 - 1^2} + \sqrt{R^2 - 2^2} + \sqrt{R^2 - 3^2} + \ldots + \sqrt{R^2 - R^2}}{R + R + R + R + \ldots + R}$$

Wallis's goal in the *Arithmetica infinitorum* is to calculate this ratio as *n*
increases to infinity, and this proves to be no easy task. He approaches
the result through a succession of approximations of similar series, that
draw ever closer to the desired ratio. But far more significant than Wallis's calculation of the area of a circle is his method for summing infinite
series that ultimately lead to his final result.

Suppose, he suggests at the start of *Arithmetica infinitorum*, that we
have a "series of quantities in arithmetic proportion, continually increasing, beginning from a point or 0 . . . thus as 0, 1, 2, 3, 4, etc." What,

he asks, is the ratio of the sum of the terms of the series to the sum of an equal number of the largest term? Wallis decides to try it out. He begins with the simplest case, of the two-term series 0, 1. The ratio is, accordingly,

$$\frac{0+1}{1+1} = \frac{1}{2}$$

He tests other cases:

$$\frac{0+1+2=3}{2+2+2=6} = \frac{1}{2}$$

$$\frac{0+1+2+3=6}{3+3+3+3=12} = \frac{1}{2}$$

$$\frac{0+1+2+3+4=10}{4+4+4+4+4=20} = \frac{1}{2}$$

$$\frac{0+1+2+3+4+5=15}{5+5+5+5+5+5=30} = \frac{1}{2}$$

$$\frac{0+1+2+3+4+5+6=21}{6+6+6+6+6+6+6=42} = \frac{1}{2}$$

Every case yields the same result, and Wallis draws a definite conclusion: "If there is taken a series of quantities in arithmetic proportion (or as the natural sequence of numbers) continually increasing, beginning from a point or 0, either finite or infinite in number (for there will be no reason to distinguish), it will be to a series of the same number of terms equal to the greatest, as 1 to 2."

Wallis could easily have proven this simple result by giving the general formula for the sum of the sequence of natural numbers beginning with 0, and dividing it by the sum of an equal number of the largest term: $\frac{n(n+1)}{2}$ divided by $n(n + 1)$, which immediately gives $\frac{1}{2}$. But his goal was not to calculate the ratio, but to demonstrate the usefulness of the method of induction: try one case, then another, and then another. If the theorem holds in all cases, then to Wallis it is proven and true. "*Induction,*"

he wrote many years later, is "a very good Method of *Investigation* . . .
which doth very often lead us to the early discovery of a General Rule."
Most important, "it need not . . . any further Demonstration."

Once he had established this first theorem, Wallis moved on to do
the same for more complex series: what if, instead of adding up a sequence of the natural numbers and dividing the sum by the same number of the largest term, he added up the squares of the natural numbers
and divided the sum by an equal number of the largest square? Using
his favored method of induction, he tries it out. Starting with the simplest case, he gets:

$$\frac{0+1=1}{1+1=2} = \frac{1}{2} = \frac{1}{3} + \frac{1}{6}$$

He then adds more terms, calculating the sum in each case:

$$\frac{0+1+4=5}{4+4+4=12} = \frac{5}{12} = \frac{1}{3} + \frac{1}{12}$$

$$\frac{0+1+4+9=14}{9+9+9+9=36} = \frac{14}{36} = \frac{1}{3} + \frac{1}{18}$$

$$\frac{0+1+4+9+16=30}{16+16+16+16+16=80} = \frac{30}{80} = \frac{3}{8} = \frac{1}{3} + \frac{1}{24}$$

$$\frac{0+1+4+9+16+25=55}{25+25+25+25+25+25=150} = \frac{55}{150} = \frac{11}{30} = \frac{1}{3} + \frac{1}{30}$$

$$\frac{0+1+4+9+16+25+36=91}{36+36+36+36+36+36+36=252} = \frac{91}{252} = \frac{13}{36} = \frac{1}{3} + \frac{1}{36}$$

Looking at the different cases, Wallis deduces that the more terms there
are in the series, the closer the ratio approaches $\frac{1}{3}$. For an infinite series,
he concludes, the difference will vanish entirely. He writes it up in a
theorem (proposition 21):

If there is proposed an infinite series, of quantities that are as squares
of arithmetic proportionals (or as a sequence of square numbers) con-

tinually increasing, beginning from a point of 0, it will be to a series of the same number of terms equal to the greatest as 1 to 3.

Wallis's proof requires only a single sentence: the result, he writes, is "clear from what has gone before." Induction needs no further support.

Wallis tries out one more series of this type, looking at the cubes rather than the squares of the natural numbers:

$$\frac{0+1=1}{1+1=2} = \frac{2}{4} = \frac{1}{4} + \frac{1}{4}$$

$$\frac{0+1+8=9}{8+8+8=24} = \frac{3}{8} = \frac{1}{4} + \frac{1}{8}$$

$$\frac{0+1+8+27=36}{27+27+27+27=108} = \frac{4}{12} = \frac{1}{4} + \frac{1}{12}$$

$$\frac{0+1+8+27+64=100}{64+64+64+64+64=320} = \frac{5}{16} = \frac{1}{4} + \frac{1}{16}$$

$$\frac{0+1+8+27+64+125=225}{125+125+125+125+125+125=750} = \frac{6}{20} = \frac{1}{4} + \frac{1}{20}$$

$$\frac{0+1+8+27+64+125+216=441}{216+216+216+216+216+216+216=1512} = \frac{7}{24} = \frac{1}{4} + \frac{1}{24}$$

The method of induction proves itself once again. As the number of terms increases, the ratio approaches ever more closely to $\frac{1}{4}$, which leads to proposition 41:

> If there is proposed an infinite series of quantities that are as cubes of arithmetic proportionals (or as a sequence of pure numbers) continually increasing, beginning from a point or 0, it will be to a series of the same number of terms equal to the greatest as 1 to 4.

Like the theorems that preceded it, this, too, requires no proof beyond self-evident induction.

In modern notation, Wallis's three theorems would look like this:

$$\lim_{n\to\infty} \frac{0+1+2+3+...+n}{n+n+n+n+...+n} = \frac{1}{2}$$

$$\lim_{n\to\infty} \frac{0^2+1^2+2^2+3^2+...+n^2}{n^2+n^2+n^2+n^2+...+n^2} = \frac{1}{3}$$

$$\lim_{n\to\infty} \frac{0^3+1^3+2^3+3^3+...+n^3}{n^3+n^3+n^3+n^3+...+n^3} = \frac{1}{4}$$

Wallis considers these ratios important steps on the road to calculating the area of a circle, because each algebraic ratio corresponded for him to a particular geometrical case. The first one shows the ratio between a triangle and its enclosing rectangle, just as Wallis shows in his proof of the area of the triangle in *De sectionibus conicis*. The series 0, 1, 2, 3, . . . , n represents the lengths of the parallel lines that make up the triangle, and the series n, n, n, n, . . . , n represents an equal number of parallel lines composing the enclosing rectangle. The ratio between them, $\frac{1}{2}$, is indeed the ratio between the areas of the triangle and rectangle (see Figure 9.1). The second case corresponds to the ratio between a half-parabola and its enclosing rectangle or, more precisely, the ratio between the area outside the half-parabola and the rectangle. The parallel lines composing this area increase as squares, that is, 0, 1, 4, 9, . . . , n^2, whereas the rectangle is represented by n^2, n^2, n^2, . . . , n^2. Wallis, in effect, shows that the ratio between the area outside a parabola and the area of its enclosing rectangle is $\frac{1}{3}$. The third ratio (Figure 9.3) corresponds to a steeper "cubic" parabola, showing that the ratio here is $\frac{1}{4}$. While Wallis still has a long way to go before calculating the more difficult ratio between the quarter-circle and its enclosing square, his strategy for arriving there is clearly taking shape.

With these results established, Wallis now makes use of induction once again to arrive at an even more general theorem: what is true for natural numbers, their squares, and their cubes, must be true for all powers m of natural numbers:

$$\lim_{n\to\infty} \frac{0^m+1^m+2^m+3^m+...+n^m}{n^m+n^m+n^m+n^m+...+n^m} = \frac{1}{m+1}$$

Wallis does not quite write the results in this form. Lacking our modern notation, he uses a table, where he assigns a ratio, $\frac{1}{2}$, to the "first power," another ratio, $\frac{1}{4}$, to the "second power," another ratio, $\frac{1}{3}$, to the "third power," and so on. The table is open-ended, and the rule is manifest: for any power m, the ratio will be $\frac{1}{m+1}$.

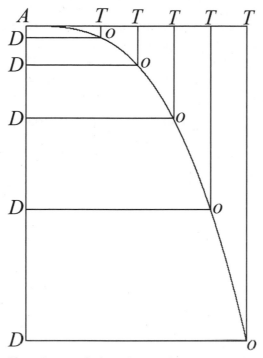

Figure 9.3. Half a cubic parabola and its enclosing rectangle. Wallis's ratios show that the ratio between the area AOT outside the cubic parabola and the area of the enclosing rectangle is $\frac{1}{4}$. After Wallis, *Arithmetica Infinitorum*, prop. 42.

Wallis viewed geometrical figures as material things, and therefore believed that, just like any object, they were composed of fundamental parts. Plane figures were made up of indivisible lines arranged next to each other, and solids of planes stacked on top of each other, just as they were for Cavalieri and Torricelli before him. But unlike the Italian masters, Wallis's preferred method for investigating mathematical objects was Baconian induction, which made his methodology resemble that of an experimentalist in his laboratory rather than a mathematician at his desk. Material, infinitesimal, and experimental, Wallis's method was one of the most unorthodox ventures in the history of Western mathematics.

It should come as no surprise, then, that not everyone was impressed with Wallis's accomplishment. Pierre de Fermat is remembered today mostly as the author of Fermat's Last Theorem, one of the longest-standing unsolved problems in mathematics, until it was proven by British mathematician Andrew Wiles in 1994. But in his day, the Frenchman was one of the most renowned and respected mathematicians in Europe. He read the *Arithmetica infinitorum* shortly after its publication in 1656, and by the following year he was engaged in a lively debate with Wallis. Fermat was skeptical, and his critiques went straight to the unconventional heart of Wallis's approach. First, he went after Wallis's infinitesimalism, which uncritically assumed that one can sum up the lines in a plane figure to calculate its area. Wallis, Fermat argued, had it backward: one cannot sum up the lines of a figure unless one already knows the area of the figure, arrived at by traditional means. If Fermat was right, then Wallis's entire project was pointless, since it pretended to demonstrate what in fact it already assumed.

If Fermat was unhappy about Wallis's casual use of infinitesimals, he was no happier about Wallis's unusual method of proof. Initially, commenting on Wallis's work to the Catholic English courtier Kenelm Digby, he was at least superficially chivalrous: "I have received a copy of the letter of Mr. Wallis, whom I esteem as I must," he began, leaving open the question of exactly how much esteem that is. Judging by what follows, it may not have been much: "But his method of demonstration, which is founded on induction rather than on reasoning in the style of Archimedes, may be somewhat difficult for novices who want demonstrative syllogisms from beginning to end." You and I, he suggested rather patronizingly, surely understand Wallis's unusual method, but mathematical "novices" might have trouble with it, and perhaps Wallis would be so good as to accommodate them. But politeness and condescension aside, it immediately became clear that Fermat's concern wasn't really the needs of the mathematically unlettered, but Wallis's method itself: it is much better, he wrote, "to prove things by the ordinary, legitimate, and Archimidean way." Wallis's method, one is left to conclude, was neither ordinary nor legitimate.

Fermat made clear the problem with induction in a separate letter, which he penned shortly thereafter. One must be extremely careful using this method, he warned, because it allows one to propose a rule that "will be good for several particulars, and nevertheless will be false in effect and not universal." The method can be useful in some circumstances, he continued, if used with care. It must not, however, "be used for the foundation of a science, that from which it is deduced, as does Mr. Wallis: for that, one must settle for nothing less than a demonstration." The unavoidable implication that Wallis's inductions were *not* demonstrations is left unsaid.

Wallis was unmoved. His mathematics of infinites, he replied, was founded on Cavalieri's method of indivisibles, and Fermat's criticism regarding the composition of geometrical figures was therefore fully answered in Cavalieri's books. Far from being a radical departure from traditional practices, his method was simply a shorthand for the irreproachable method of exhaustion used by the ancient masters Eudoxus and Archimedes. If Fermat nevertheless wished to reconstruct all the proofs in the classical form, Wallis wrote, "it was free for him to do it." But "he might spare himself the labour, because it was already done to his hand by Cavallerius."

Wallis artfully deflected Fermat's valid criticism of infinitesimals without answering it directly. The claim that "there is nothing new here" sounds disingenuous coming from someone who loudly proclaimed the novelty of his work. "You may find this work new (if I judge rightly)," he wrote in his dedication of the *Arithmetica infinitorum* to William Oughtred, adding that "I see no reason why I should not proclaim it." His claim that Cavalieri had already answered all objections was an effective strategy also used by Torricelli, Angeli, and other promoters of the infinitely small. It ignored the withering attacks on Cavalieri by the Jesuits and others, thereby presenting indivisibles as far more widely accepted than they actually were. It also relied on the high likelihood that Fermat had never actually read Cavalieri's tomes, whose notorious unreadability provided cover to many seventeenth-century indivisiblists.

Wallis was equally unimpressed by Fermat's critique of his method of induction. Proofs by induction, Wallis claimed, "are plain, obvious and easy," and require no additional demonstration. "If any think them less valuable," he wrote, "because not set forth with the pompous ostentation of Lines and Figures, I am quite of another mind." Any competent mathematician who put in the time, Wallis argued, could convert his proofs by induction into traditional geometric proofs, but to do so would be mere fussiness: "I do not find that *Euclide* was wont to be so pedantick," he wrote, and "I am sure *Archimedes* was not." Pedants such as Fermat, according to Wallis, were a distinct minority: "[M]ost mathematicians that I have seen, after such induction continued for some steps . . . are satisfied (from such evidence) to conclude universally *and so in like manner for the consequent powers*. And such Induction hath been hitherto thought . . . a conclusive argument." With these brief and contemptuous remarks, Wallis dismissed thousands of years of tradition.

WALLIS SAVES MATHEMATICS

If Wallis's approach was unacceptable to comparatively orthodox mathematicians such as Fermat, for his colleagues at the Royal Society it was the solution to a vexing problem. Boyle, Oldenburg, and their associates had enshrined the experimental method as the proper way to pursue science. To them, it was not only the correct methodology for revealing the secrets of nature, but also a model for the proper workings of the state. Unfortunately, while experimentalism supported the Royal Society's founders' vision of both nature and society, it also left mathematics on the wrong side of the methodological and political divide. Mathematics, as commonly understood, left no room for competing points of view, extracting agreement through the irresistible power of its reasoning rather than reaching agreement through freely given consent. It was the exclusive domain of a small number of experts whose work was too technical and esoteric to be competently evaluated by intelligent and educated laymen, and whose pronouncements—absolute and arrogant—had to be accepted as true based on their authority alone. To top it off, mathematics was the cornerstone of a vision of knowledge and of the state that the Royal Society's grandees viewed with disgust and horror: Hobbes's authoritarian science and totalitarian commonwealth. As they saw it, whereas experimentalism stood for moderation, tolerance, and peace, mathematics was the tool of the advocates of dogmatism, intolerance, and their inevitable outcome, civil war.

This left the founders of the Royal Society in a quandary. How could they retain the power and scientific accomplishments of mathematics without also taking on board its unwelcome baggage? Wallis had the answer: his unique brand of mathematics was as powerful as the traditional approach, but also perfectly in line with the Society's cherished experimentalism. To the founders of the Royal Society it was a godsend: here was a flexible mathematical approach that accommodated dissenting views and was modest in its claims. Precisely the kind of mathematics the Royal Society could endorse, and promote.

To see just how different Wallis's mathematics was from the rigid Euclidian approach detested by the Society, it is instructive to compare his practice with the mathematical views of the man the Society feared most, Thomas Hobbes. To begin with, Hobbes insisted that geometrical entities must be constructed by us, from first principles, and were consequently fully known. Not so, retorted Wallis: lines, planes, and geometrical figures were given to us fully formed, and their mysteries should be investigated just as a scientist studies natural objects. Then came the issue of mathematical method, with Hobbes insisting that strict deduction was the only acceptable way to proceed in mathematics, since it alone secured absolute certainty. Wallis, in contrast, advocated induction, which, he argued, was far more effective than deduction in discovering new results. The fact that induction never pretended to achieve the level of certainty that Hobbes so prized was for Wallis a small price to pay. Finally, since Hobbes insisted that his mathematical deductions arrived at absolute truth, he cared nothing for the opinions of others. The proof stood for itself, whether others understood it or not. But Wallis's inductive proofs are not infallible logical deductions, but rather, strong, persuasive arguments aimed at swaying his audience. Their success depended very much on whether Wallis's readers believed in the end that the theorems were true for all cases, not just the ones he tried out.

In almost every aspect, Wallis's mathematics replicated the experimental practices of his associates at the Royal Society. He investigated external objects, not constructed ones; his mathematics relied on induction, not deduction; it never claimed to arrive at a final truth; and the ultimate arbiter of this truth was the consensus of men. It was precisely the kind of mathematics that one would expect from the only mathematician among the founders of the Royal Society, and it was precisely what the Society grandees were looking for. Instead of being a dangerous rival to experimental practices, mathematics could now join with them to promote proper science and a proper political order.

Wallis and Hobbes both believed that mathematical order was the foundation of the social and political order, but beyond this common assumption, they could agree on practically nothing else. Hobbes advocated a strict and rigorous deductive mathematical method, which was his model for an absolutist, rigid, and hierarchical state. Wallis advocated a modest, tolerant, and consensus-driven mathematics, which was designed to encourage the same qualities in the body politic as a whole. Across the mathematical and political divide the two faced each other, and the stakes could not have been higher: the nature of truth; the social and political order; the face of modernity.

GOLIATH AGAINST THE BACKBITERS

The first volley in the war between the Savilian Professor of Geometry and the courtly political philosopher was fired in the summer of 1655, when Wallis published the *Elenchus geometriae Hobbianae*, a scathing critique of Hobbes's geometrical efforts in *De corpore*. The last volley was fired twenty-three years later, when the ninety-year-old Hobbes published *Decameron physiologicum*, which included a discussion of "the proportion of a straight line to half the arc of a quadrant." It was Hobbes's last effort to defend his mathematics and undermine his rival, but the back-and-forth would likely have continued indefinitely had Hobbes not died the very next year. In between, Wallis published an additional ten books and essays aimed directly at Hobbes, while Hobbes published at least thirteen tracts aimed specifically at Wallis. To these may be added innumerable other insults, slurs, accusations, and (occasionally) serious critiques, which constituted asides in other works by these two very prolific authors. When the battle was at its height, accusations were flying back and forth at a ferocious pace, with each man charging the other with not only mathematical incompetence, but also political subversion, religious heresy, and personal villainy.

When the war began, the two had very likely never met. Wallis undoubtedly knew of the celebrated author of *Leviathan*, and his friend and colleague Seth Ward had traveled to London to meet Hobbes when he returned to England in 1651. If Hobbes knew of Wallis at all, it was only as the new (and apparently unqualified) Savilian Professor at Oxford, appointed by Parliament for overtly political reasons. Even in later years, when the two spent a great deal of their time trying to demolish each other's reputations, there is no record that they ever met, although it is hard to imagine that they did not run into each other in the tight social circles of the English intellectual elite. The conflict between them was rooted in their opposing political, religious, and methodological views, not in personal animosity. But it did not take long for the exchanges to turn personal, and vicious.

Wallis set the tone early on: "No-one can doubt how puffed up with pride and arrogance is this man," he wrote in his dedication of *Elenchus* to John Owen, dean of Christ Church College and vice chancellor of the University of Oxford. "When I look at him, the equal of 'Leviathan' (which made his name for him) or rather 'Goliath,' parading with such arrogance, I decided that he should be thoroughly attacked, so that he may see he cannot do anything he likes without being called to task . . ." A puffed-up and arrogant Goliath parading around as if he had sole possession of the truth became Wallis's favored caricature for Hobbes, and he vowed to "burst the balloon" of "that man, so full of airy talk." As for Hobbes, he seemed unfazed by Wallis's abusive language, and while occasionally complaining about his opponent's uncivil tone, he joined in the war with relish and gave as good as he got.

The condescending title of Hobbes's first response to Wallis's invective foreshadowed much that was to follow: *Six Lessons to the Professors of Mathematics, One of Geometry, the Other of Astronomy*. If Wallis considered Hobbes arrogant before, then this tract surely confirmed it. Here was Hobbes, a house intellectual for the Cavendish clan, with no credentials or position, presuming to teach geometry to Wallis and Ward, holders of two of the most distinguished mathematical chairs in Europe. Nor did Hobbes stop there, for in his dedication to Lord Pierrepont, he went on to argue that he was in fact far more deserving of their positions than they: by setting forth the true foundations of geometry in *De corpore*, he argued, "I have done that business for which Dr. Wallis receives the wages."

In the work itself, Hobbes moves from defending his own mathematical work to deriding Wallis, and answers contempt with contempt: "I verily believe," he wrote of the *Arithmetica infinitorum* and Wallis's work on the angle of contact, "that since the beginning of the world there has not been, nor shall be, so much absurdity written in geometry as is to be found in those books." Of Wallis's use of algebraic symbols, some of them (such as ∞) of his own invention, he opines that "symbols are poor, unhandsome, though necessary, scaffolds of demonstration; and ought no more to appear in public, than the most deformed necessary business which you do in your chambers." Wallis's "On Conic Sections," according to Hobbes, "is so covered with the scab of symbols, that I had not the patience to examine whether it be well or ill demonstrated." It may be that Wallis had these witticisms in mind when he complained, years later, of those who deride symbols and insist on classical proofs decked in "the pompous ostentation of Lines and Figures." When Wallis tried to respond to some of Hobbes's more substantial criticisms, Hobbes brushed him off like an imperious schoolmaster disciplining an unruly child: "You do shift and wiggle," he wrote impatiently, "and throw out ink, that I cannot perceive which way you go, nor need I." But it is all to no avail: "your book of the *Arithmetica Infinitorum* is all nought, from the beginning to the end."

And so it went, back and forth, for nearly a quarter century. Hobbes was the better stylist and sharper wit, but Wallis held his own thanks to the sheer fervor and volume of his denunciations. He was also by far the better connected of the two, using his positions at Oxford and the Royal Society to gradually isolate Hobbes and discredit him among the English literati. If in the 1650s Hobbes was widely considered a formidable scientist and mathematician, by the 1670s he was seen as a political philosopher who had unwisely strayed beyond his field of expertise, only to be exposed as an incompetent amateur. Even Charles II, Hobbes's former student, joined the general fun of tormenting the aging philosopher: "Here comes the beare to be bayted!" the king would announce when

Hobbes made one of his frequent visits to court. Consequently, despite being one of the most celebrated scholars in England, and despite having many acquaintances in the ranks of the Royal Society, the aging philosopher was never elected a fellow. Hobbes attributed this to the implacable hostility of his powerful enemies in the Society, Wallis and Boyle, and there is undoubtedly some truth to this. But there is also the fact that his bruising decades-long fight with Wallis (and a shorter fight with Boyle) left his scientific reputation in tatters, so that he could legitimately be dismissed as unworthy of being a fellow.

But beneath the sound and fury generated when two intellectuals engage in public combat, much was in fact at stake. Wallis said as much when he explained why he had launched his first assault on Hobbes's mathematics. Why, he asked, "should I undertake to refute his Geometry, leaving out Theology and other Philosophies, when there are other things in which he has made far more dangerous errors?" The reason, he explains, is that Hobbes had "set such store by geometry that without it there is hardly anything sound that could be expected in philosophy." Hobbes is so sure of the mathematical grounding of his system that "if he sees anyone disagreeing with him in Theology or Philosophy he thinks they should be sent away with the supercilious reply that since they are unlearned in geometry they do not understand those things." The one sure way to topple Hobbes's entire philosophical edifice is to show that he is in fact a mathematical ignoramus. Then the man "so full of airy talk" will be "quite deflated," and people will know ". . . that there is no more to be feared of this Leviathan on this account, since his armour (in which he had the greatest confidence) is easily pierced."

Wallis repeated this explanation, after enduring several punishing rounds with Hobbes, in a 1659 letter to the Dutch polymath Christiaan Huygens. The "very harsh diatribe" against Hobbes, he explained, was not caused by his lack of manners, but by the "necessity of the case." It was "provoked by our Leviathan, when he attacks with all his might, and destroys our universities . . . and especially the clergy, and all institutions and all religion." Since this "Leviathan" relied so much on mathematics, Wallis continued, "it seemed necessary that now some mathematician . . . should show how little he understands this mathematics (from which he takes his courage)." Destroying Hobbes's mathematical credibility would discredit his teachings and preserve the institutions threatened by his destructive philosophy.

Fortunately for Wallis, Hobbes's unconventional brand of geometry offered many openings for attack by a skilled mathematician. If Hobbes had stuck close to the classical geometry that had captivated him years before, when he stumbled upon an open volume of Euclid in a continental gentleman's library, he would have been on surer ground. But for Hobbes this was not enough: in order to support his political system, his geometry must be a perfect science, capable of resolving all outstanding problems. And in his quest to transform geometry into this ideal, Hobbes foundered. This is not because he was a mathematical ignoramus, since his attempted proofs show a powerful mathematical mind at work. It was because it is simply impossible to resolve the ancient problems by classical means, and the project he had embarked upon was doomed from the start.

Up to a point, Hobbes was following in the footsteps of Clavius and the Jesuits, who tried to use geometry as a model for the proper order of knowledge, society, and the state. The rigid state they advocated, in which the word of the sovereign (for Hobbes) or the Pope (for the Jesuits) had the force of law, and all opposition was deemed absurd, mirrored the rational geometrical order. But Hobbes went a step further than the Jesuits: rather than settling on treating geometry as a model and an ideal, he tried logically and systematically to derive his philosophy from his modified geometrical principles. This required that he show that all things in the world could be constructed from geometrical principles, and in *De corpore*, he set out to do just that.

But the world, as it turns out, cannot be derived from mathematics. The Pythagoreans learned this more than two thousand years earlier, when the existence of incommensurable magnitudes upended their belief that everything in the world could be described in terms of the ratio of whole numbers.

Hobbes tried to defend himself. When Wallis ridiculed his repeated failures at squaring the circle, he protested about Wallis's regurgitation of discarded results: "Seeing you knew I had rejected that Proposition," Hobbes wrote, "it was but poor Ambition to take wing as you thought to do, like Beetles from my egestions." He admitted his mistakes in his first two attempts in *De corpore*, but insisted that the cause of the error was mere negligence, and not any problem with his method.

Hobbes would not, and probably could not, concede that his fundamental approach was flawed. His entire philosophical system was at stake, he believed, and to admit that his geometry was hopeless was, to him, the same as admitting that everything he had ever written down and argued for was worthless. So, repeatedly, for more than two decades, Hobbes kept producing new "proofs," and Wallis, who well understood how critical they were for Hobbes, kept demolishing them. Cornered and isolated, and facing a rising tide of mathematical criticism, Hobbes hunkered down. "I do not wish to change, confirm, or argue anymore about the demonstration that is in the press," he wrote to his friend Sorbière in 1664 about yet another of his quadratures of the circle. "It is correct; and if people burdened with prejudice fail to read it carefully enough, that is their fault, not mine." He went to his grave fully convinced that he had succeeded in squaring the circle.

WHICH MATHEMATICS?

Wallis had his fun demolishing each of Hobbes's solutions to the three classical problems whenever the old philosopher proposed a new one. But he also had methodological criticisms of his rival's mathematical approach, and these too made their way into his denunciations of Hobbes's mathematics. In particular, Wallis objected to Hobbes's attempt to construct mathematics from physical material principles. "Who ever, before you, defined a point to be a body? Who ever seriously asserted that mathematical points have any magnitude?" If points have size, Wallis continued, then adding two, three, or a hundred points together will increase their size two, three, or a hundred times, which is absurd. The same critique applied to Hobbes's use of other physical properties in his definition of geometrical concepts such as his "conatus" and "impetus." Hobbes needed these terms because he believed that all proofs must be driven by material causes. But Wallis, seeing an opening, adopted the classical Euclidean position that drew a strict distinction between perfect geometrical objects and their flawed material counterparts: "To what end," Wallis asked, "need there be a consideration of time or weight or any other such quantity" in geometrical definitions? Such physical attributes, he argued, have no place in the world of geometry.

Wallis's criticism of Hobbes on the grounds that his mathematics incorporated material notions was, to put it mildly, disingenuous. After all, Wallis's own mathematical approach did much the same thing: positing geometrical objects as existing "out there" in the world, dividing them up into their indivisible components, and studying them experimentally. Indeed, Hobbes was happy to return the favor by condemning Wallis for precisely that. Nevertheless, Wallis's mathematics was, in the end, far less vulnerable to methodological attacks than was Hobbes's, because Hobbes wanted to use mathematics as a bulwark of certainty that would buttress his political philosophy. Consequently, any criticism that raised questions about the logical soundness of his approach struck at the very core of his enterprise. Wallis, in contrast, cared little for methodological certainties. His purpose, he explained years later, "was not so much to shew a Method of Demonstrating things already known . . . as to shew a way of *Investigation* of finding out of things yet unknown." It was the effectiveness of his approach in establishing new results that mattered to Wallis. Using a perfect and irreproachable method of proof counted for very little indeed.

Hobbes could never accept this. Muddled foundations, he was convinced, led to muddled thinking, confused knowledge, controversy, and social strife. The only possible reason for Wallis and his fellow divines to be promoting such indefensible stances as magnitudeless points and immaterial spirits, it seemed to Hobbes, was that they were acting out of self-interest, trying to appropriate for themselves the authority that rightfully belonged to the civil sovereign. He was determined to stop them. *Leviathan*, he explained, was written to expose how much the English ministers, in their efforts to seize as much power as possible, had contributed to the outbreak of the civil war, and his struggle with Wallis was also part of the same struggle against the power-hungry clerics. "My business," he explained to Sorbière in 1655, was "with all the ecclesiastics of England at once, on whose behalf Wallis wrote against me." The way to stop the conspiracy and save the commonwealth was to expose the falsehoods of Wallis's mathematics and drive their author in disgrace from the ranks of mathematicians.

Hobbes struck directly at the two main pillars of Wallis's mathematics. First came induction: "Egregious logicians and geometricians," Hobbes exclaimed in disbelief, "that think an induction . . . sufficient to infer a conclusion universal and fit to be received for a geometrical demonstration!" The fact that a rule holds for a certain number of cases says nothing about whether it will hold for other cases that were not tested. Fermat had made the same point, as indeed would any classically trained mathematician.

But it was the second pillar of Wallis's mathematics that was the focus of Hobbes's true scorn: the concept of the infinitely small. Wallis had calculated the areas of a triangle by dividing it into an infinite number of parallel lines, each with an infinitesimal width of $\frac{1}{\infty}$, and then summing them up. Hobbes, sensing vulnerability, struck with force and precision. "In the first proposition of your *Conic Sections* you first have this, '*that a parallelogram whose altitude is infinitely little, that is to say none, is scarce anything else but a line*,'" Hobbes began, quoting Wallis's own words. "Is this the language of geometry?" he thundered. "How do you determine this word *scarce?*" Intuitively we know quite well what Wallis means by this term, but Hobbes is perfectly correct that *scarce* is not a mathematical term. This was not a minor issue: The whole point of studying mathematics, for Hobbes, was its rigor, precision, and certainty. Using ambiguous terminology, as Wallis was fond of doing, undermined the entire enterprise.

But Hobbes had more in store for Wallis. The altitude of the lines/ parallelogram that comprise the triangle is either something or nothing, and Wallis's proof founders either way. If the lines have no breadth, Hobbes charges, "the altitude of your triangle consisteth of an infinite number of no altitudes, that is of an infinite number of nothings, and consequently the area of your triangle has no quantity." Allowing for the lines to possess a certain width and form minuscule parallelograms is just as disastrous for Wallis's proof: "If you say that by the parallels you mean infinitely little parallelograms, you are never the better," Hobbes argues. This is because the opposite sides of the supposed parallelograms in Wallis's construction form the sides of the triangle. And since, as Hobbes points out, no two sides of a triangle are parallel, neither are the opposite sides of the component parallelograms. This leads inevitably to the conclusion that they are not parallelograms at all.

On other points in Wallis's *Arithmetica infinitorum*, Hobbes was even more scathing. In his proofs, Wallis calculated ratios between an increasing infinite series in the numerator, and an "equal number" of the largest term in the series. But how can an increasing infinite series have a "largest term?" And how can two infinite series, one in the numerator, the other in the denominator, have the "same number" of terms? Acting as if infinite series have a largest term or a certain number of terms, Hobbes charges, is equivalent to treating the infinite as finite, which is a contradiction in terms. "This principle is so absurd," Hobbes rails, "that I believe it could hardly have been proposed by a sane person." To Wallis's casual assertion that Cavalieri showed *"that any continuous quantity consists of an infinite number of indivisibles,* or of infinitely small parts," Hobbes responds that although he read Cavalieri's book (which he suspected rightly Wallis had not), he did not recall that it contained anything of the sort. "For it is false. A continuous magnitude is by its nature always divisible into divisible parts: Nor can there be anything infinitely small."

THE BATTLE FOR THE FUTURE

And that, indeed, was the crux of the matter: Hobbes rejected the concept of the infinitely small and the mathematics that went with it. Mathematics, he insisted, must begin with first principles, and proceed deductively, step by step, to ever-more-complex but equally certain truths. In the process, all geometrical objects must be constructed from simpler ones, using only the simple, self-evident definitions of point, line, surface, and so on. In this manner, Hobbes believed, an entire world could be constructed—perfectly rational, absolutely transparent, and fully known, a world that held no secrets and whose rules were as simple and absolute as the principles of geometry. It was, when all was said and done, the world of the Leviathan, the supreme sovereign whose decrees have the power of indisputable truth. Any attempt to tinker with the perfect rational reasoning of mathematics would undermine the perfect rational order of the state, and lead to discord, factionalism, and civil war.

As Hobbes saw it, however, the infinitely small, an unwelcome intruder into mathematics, did precisely that: it destroyed the transparent rationality of mathematics, which in turn undermined the social, religious, and political order. For one thing, it did not construct its objects logically and systematically, as mathematics must if it is to be the basis of a universal rational order. Even more critically, the infinitesimal itself was notoriously paradoxical and even self-contradictory, and could as easily be used to produce obvious errors as truths. Such a nonconstructive, paradoxical approach represented to Hobbes everything that mathematics must never be. If the infinitely small were allowed into mathematics, then all order would be at risk, and society and state would descend into ruin. In the vague and ill-defined infinitely small, Hobbes perceived an echo of the unruly Diggers on St. Georges Hill.

Wallis saw things differently. Practically all the features of the infinitely small that Hobbes considered a disaster in the making Wallis considered clear advantages. Hobbes was convinced that the very existence of dissent led inexorably to chaos and strife, and was determined to quash any hint of it. Mathematics, the only science that (he believed) had succeeded in eliminating dissent served him for this purpose. But Wallis, along with his fellows at the Royal Society, believed that it was dogmatism and intolerance that had led to the disasters of the 1640s and '50s. Their concern was not that knowledge would be uncertain, but that it would appear to be too certain and dogmatic, excluding competing beliefs. Wallis's mathematics offered an alternative.

Unlike classical Euclidean geometry, Wallis's mathematics did not attempt to construct a mathematical world, but instead to investigate the world as it was. This in itself made it far more palatable to those who feared a strict rational world order. Wallis's world was still mysterious, unexplored, and ready for new investigations, by mathematical or other means. The ambiguity of the infinitely small, far from disqualifying it as a proper mathematical concept, was also a positive feature: opaque and even paradoxical, it left room for different interpretations and explanations of its nature and workings. Finally, Wallis's infinitely small proved remarkably successful in revealing new mathematical truths, demonstrating the power of this incompletely understood concept. The way forward, it was clear to its advocates, was to proceed carefully, experimentally, gradually and laboriously, using whatever worked, to reveal the mysteries of the world. Any attempt to construct a perfectly known and rational mathematical world was not only politically dangerous, but also a scientific dead end.

Epilogue:
Two Modernities

Wallis won. By the 1660s, only a few years into a war that would last decades, Hobbes was effectively ejected from the ranks of mathematicians, while Wallis remained an honored member of the republic of mathematics. Unlike his associate and fellow Hobbes nemesis Seth Ward, he was not appointed an Anglican bishop, and it is likely that the ardent Presbyterianism of his early years prevented such an appointment. But he remained Savilian Professor and keeper of the archives at Oxford, was regularly employed by the king as a code-breaker for captured correspondence, and was a frequent visitor to court. Among his friends he counted the founders of the Royal Society, including Henry Oldenburg and Robert Boyle, and also their illustrious successors, Isaac Newton and John Locke. He kept publishing throughout his life not only on mathematics but also on mechanics, logic, and English grammar, and he considered himself an expert on teaching the deaf and mute to speak. Two collections of some of the dozens of sermons he delivered over several decades were published during his lifetime. At his death in 1703, at the age of eighty-six, he was mourned as "a man of most admirable fine parts, and great industry, whereby in some years he became so noted for his profound skill in mathematics that he was deservedly accounted the greatest person in that profession of any in his time."

But more important than his personal and professional success was that Wallis's controversial mathematics had prevailed. Whereas Hobbes's proofs were generally dismissed as the work of an amateur, Wallis's results in the *Arithmetica infinitorum* and elsewhere were checked and confirmed by his colleagues. Unquestionably the most important reader of the work was Isaac Newton. When the twenty-three-year-old Newton worked out his own version of infinitesimal mathematics in 1665, the *Arithmetica infinitorum*, he later reported, was one of his chief sources of inspiration. In the following decades, Newton's calculus, as well as its rival Leibnizian version, spread far and wide, transforming the practice of mathematics and all the mathematical sciences. Analysis, the new mathematical field that took the calculus as its starting point, became the dominant branch of mathematics in the eighteenth century, and remains one of the chief pillars of the discipline. It made possible the mathematical study of everything from the movement of the planets, to the vibrations of strings, to the workings of steam engines, to electrodynamics—practically every field of mathematical physics from that day to ours. Wallis lived to see only the very early stirrings of this mathematical revolution that transformed the world over the following centuries. But when he died in 1703, the verdict was already in: the infinitely small had won the day.

From Rome and Florence to London and Oxford, the fight over the infinitely small raged across Western Europe in the middle decades of the seventeenth century. And at the opposite ends of the Continent, the struggle yielded opposite outcomes: in Italy the Jesuits prevailed over the Galileans, whereas in England Wallis prevailed over Hobbes. These outcomes were not a foregone conclusion: if an objective observer in 1630 had been asked to predict the fortunes of mathematics in the two lands, he would almost certainly have predicted the opposite. Italy was home to an illustrious mathematical tradition, whereas England had never produced any geometers of note, with the possible exception of the reclusive Thomas Harriot, who never published. If any land was likely to pioneer a challenging new mathematics, it was Italy, whose art and science had inspired Europe since the Renaissance. England, meanwhile, would likely have remained what it had always been, an intellectual backwater feeding off the scraps of its more cultured continental neighbors.

But things turned out differently. Following the war over the infinitely small, advanced mathematics in Italy came to a standstill, whereas English mathematics quickly became one of the dominant national traditions in Europe, rivaled only by the French.

To appreciate the magnitude of the effect that the acceptance of infinitesimals had on the modern West, consider what the world would have been like without them. If the Jesuits and their allies had had their way, there would be no calculus, no analysis, nor any of the scientific and technological innovations that flowed from these powerful mathematical techniques. As early as the seventeenth century the "method of indivisibles" was applied to problems of mechanics, and proved particularly effective in describing motion. Galileo used it to describe the motion of falling bodies, and his contemporary Johannes Kepler used it to discover and describe the movement of the planets around the sun. Isaac Newton made use of the calculus to create a new physics and mathematically describe the complete "system of the world" bound together by universal gravitation.

Newton's work was followed up in the eighteenth century by luminaries such as Daniel Bernoulli, Leonhard Euler, and Jean d'Alembert, who offered general mathematical descriptions of the motion of fluids, the vibrations of strings, and the currents of the air. Their successors Joseph-Louis Lagrange and Pierre-Simon Laplace were able to describe the mechanics of both heaven and earth in a set of elegant "differential equations"—that is, equations that make use of the calculus. Indeed, from their day to ours, analysis—the broader form of the calculus—has remained the fundamental tool used by physicists to explain natural phenomena. And all this had its roots in the "indivisible Line of the Mathematicians" that troubled Sorbière in 1664.

The impact of the calculus on engineering and technology took a little longer to develop, but when it did, it was just as revolutionary. The mathematical theory of heat diffusion developed by Joseph Fourier and thermodynamics developed by William Thomson in the nineteenth century made possible the design and production of ever-more-efficient steam engines. In the 1860s, James Clerk Maxwell wrote down what became known as the Maxwell equations, a set of differential equations describing the relationship between electricity and magnetism. The subsequent development of electric motors and generators and of radio communications would never have been possible without his work. Add to that the fundamental role of the calculus in aerodynamics (making possible air travel), hydrodynamics (shipping, water collection and distribution), electronics, civil engineering, architecture, business models, and on and on, and the picture becomes clear: the modern world would have been unimaginable without the calculus and the insights it opened up into the workings of the natural world.

But the story does not end there, for the implications of the struggle over the infinitely small ranged far beyond the mathematical world, or even that of science and technology. The fight was over the face of modernity—and indeed, in the two lands in which the struggle raged fiercest, modernity took starkly different paths. Italy became very much what the Jesuits had made it. Profoundly Catholic, it was imbued with the eternal, unchanging truths of Catholic dogma and dominated by the absolute spiritual authority of Pope and Church hierarchy. This spiritual order supported a secular order that shared many of the same characteristics. While papal supremacy did not allow for the development of a strong state, the petty princelings who ruled Italy as tyrannical autocrats owed their positions to ancient dynastic rights and claimed absolute authority over their domains. Religious dissent was unthinkable, political opposition fiercely repressed, intellectual innovation frowned upon, and social mobility nearly nonexistent. As nations to the north became hotbeds of intellectual debate, technological innovation, political experimentation, and economic progress, Italy remained frozen in time. The land that for centuries had led all Europe in the arts and sciences, whose famous city-states were at one point wealthier and more prosperous than many a great kingdom, became stagnant, backward, and poor.

In the same years that Italy was falling behind, England was becoming the most dynamic, forward-looking, and fastest-growing land in Europe. Long considered a wild and semibarbarous country at the northern edge of European civilization, it became a leading center of European culture and science and a model for political pluralism and economic success. Here was a vision of modernity opposite in every way to the one that dominated Italy: Instead of dogmatic uniformity, England exhibited a marked and ever-increasing openness to dissent and pluralism. Politically, religiously, and economically, England became a land of many voices, where rival views and interests competed openly, relatively free from state repression. And it was in this relative freedom that England discovered its path to wealth and power.

The efforts of the Stuart kings to establish an absolute monarchy were defeated by fierce resistance from Parliament, first during the civil

war and finally in the Glorious Revolution of 1688. When the last of the
Stuarts, James II, was driven into exile and replaced by the constitutionally
minded William of Orange, Parliament became the supreme governing
body of the state. To be sure, the English Parliament of the seventeenth
and eighteenth centuries was very far from the democratic institution
we know today. It was a conservative body that represented the landed
and propertied classes, and feared social unrest by the landless far more
than it feared royal domination. Nevertheless, it was a deliberative body
that allowed for an unprecedented degree of dissent, debate, and the
free expression of ideas. Over time, the openness inherent in a parliamentary system triumphed over the class and social loyalties of its members.
Throughout the eighteenth and nineteenth centuries the franchise was
slowly but irreversibly expanded, and Parliament's membership expanded
to include ever broader segments of the population. The process was not
completed until 1928, when all women were given the right to vote.

Political pluralism in England was matched by an unprecedented
degree of religious tolerance. The Puritans of the early seventeenth century were not a tolerant lot. They viewed themselves as God's elect, and
their zeal to impose their beliefs and morals on the broader population
played no small part in launching the crisis of 1640 and the civil war
that followed. But the tragedy of the civil war, broadly attributed to the
clash of competing dogmas, gave rise to a more forgiving attitude toward
religious truth. Many of the Anglican bishops in the aftermath of the
Interregnum, and nearly all the leaders of the newly founded Royal Society, advocated a latitudinarian, rather than dogmatic, approach to religion. Instead of insisting on a strict religious doctrine and excluding
anyone who did not fully subscribe to it, they advocated broad latitude
in doctrinal matters, acknowledging that ultimate truth was something
to search for, not a given. A range of different beliefs was welcomed
within the Anglican fold, as long as those beliefs agreed on certain fundamental tenets, such as the Trinity and the supremacy of the king
(rather than the Pope).

First advocated for the Church of England itself, religious pluralism
was soon extended beyond the Church's confines. The 1689 Act of Toleration, coming fast on the heels of the Glorious Revolution, granted
freedom from persecution to nonconforming Protestant sects such as
Presbyterians, Quakers, and Unitarians. Though restricted (until 1828)

from many spheres of public life, and from the universities of Oxford and Cambridge, the dissenters were nevertheless secure from state interference and prospered both economically and intellectually. They formed their own churches and their own academies, which were often far more advanced in their teachings than the plodding Anglican universities. Catholicism was not as easily tolerated, however. Both detested and feared, it was associated with the danger of foreign intervention and papal claims of supremacy, and with the claims of the deposed Stuarts. But even though they suffered systematic discrimination, English Catholics were largely left in peace until their official emancipation in 1829.

Political and religious pluralism went hand in hand with scientific, intellectual, and economic openness. The Royal Society, along with the French Academy of Sciences, soon became the leading scientific academy in Europe, and English science set the standard for all Europe. In the realm of letters, England became a locus of public philosophical and political debates in which luminaries such as John Locke, Jonathan Swift, and Edmund Burke took opposing, brilliantly argued, sides. Political liberalization also made possible economic liberalization and private entrepreneurship on an unprecedented scale. The accumulation of capital and the growing size of workshops made it profitable to invest in new technologies, particularly the steam engine. As a result, by the late eighteenth century, England became the first industrialized country in the world, pulling far ahead of its continental rivals and leaving them scrambling to keep up.

Whether the continuum is made up of infinitesimals seems like the quaintest of questions, and it is hard for us to fathom the passions it unleashed. But when the struggle raged in the seventeenth century, the combatants on both sides believed that the answer could shape every facet of life in the modern world that was then coming into being. They were right: when the dust cleared, the champions of infinitesimals had won, their enemies defeated. And the world was never the same again.

DRAMATIS PERSONAE

The "Infinitesimalists"

Luca Valerio (1553–1618): A mathematician and friend of Galileo who made important contributions to infinitesimal methods. However, when Galileo clashed with the Jesuits in 1616 Valerio sided against him, and was fiercely denounced by his former friends. He died in disgrace shortly thereafter.

Galileo Galilei (1564–1642): The most celebrated scientist of his day. He was persecuted by the Jesuits for his advocacy of Copernicanism, which led to his trial and downfall. Galileo made use of infinitesimals in his work, and supported and encouraged a generation of young mathematicians to develop the concept. Even after his condemnation, he was still the undisputed leader of the Italian infinitesimalists.

Gregory St. Vincent (1584–1667): A Jesuit mathematician who developed a novel method for calculating the volumes of geometrical figures that involved infinite division. His Jesuit superiors considered the method too close to infinitesimals, and forbade him from publishing his work.

Bonaventura Cavalieri (1598–1647): Galileo's disciple, later professor of mathematics at the University of Bologna, and a member of the order of the Jesuats. His books *Geometria indivisibilibus* (1635) and *Exercitationes geometricae sex* (1647) became the standard works on the new mathematics, which he called "the method of indivisibles."

Evangelista Torricelli (1608–47): Galileo's disciple and ultimately his successor in Florence. An ardent infinitesimalist, and far less concerned with technical rigor than Cavalieri, he was famous for his powerful and creative techniques, which involved calculating the "width" and "thickness" of infinitesimals. His *Opera geometrica* of 1644 was widely read by mathematicians across Europe, and Wallis in particular modeled his *Arithmetica infinitorum* on Torricelli's work. Among Torricelli's most surprising results was his success in calculating the volume of a solid of infinite length.

John Wallis (1616–1703): An ardent Parliamentarian and Puritan divine in the early years of the Interregnum, Wallis served as secretary to the Westminster Assembly of Divines. From the mid-1640s he was a regular participant in the private meetings that would later lead to the establishment of the Royal Society of London, and in 1649 he was appointed Savilian Professor of Geometry at the University of Oxford. Wallis made his name in mathematics as a leading infinitesimalist, and in politics as a relative pragmatist and moderate, in line with his fellows at the Royal Society. He was engaged in a decades-long war with Hobbes over his mathematics and his authoritarian politics.

Stefano degli Angeli (1623–97): A friend and disciple of Cavalieri, professor of mathematics at the University of Padua, and a member of the order of the Jesuats. In the 1650s and '60s he was the last public voice in Italy defending infinitesimals and openly denouncing the Jesuits. But when the Jesuats were unceremoniously dissolved by the Pope in 1668, Angeli finally desisted, and never published on infinitesimals again.

The "Anti-Infinitesimalists"

Christopher Clavius (1538–1612): Professor of mathematics at the Jesuit Collegio Romano and founder of the Jesuit mathematical tradition. Clavius insisted on a geometrical approach, which he valued for its orderliness, rigorous deductive method, and absolutely true results. He hoped to apply this methodology to all fields of knowledge, and was not interested in mathematical innovation. Clavius did not address infinitesimals directly, since they were hardly used by mathematicians during most of his career, but he was the author of the core principles of Jesuit mathematics, which led directly to the war on infinitesimals.

Paul Guldin (1577–1643): Leading Jesuit mathematician, charged with discrediting infinitesimals. He attacked Cavalieri's method in his *De centro gravitatis* of 1641.

Mario Bettini (1584–1657): Mathematician who became the leading Jesuit critic of the infinitely small after Guldin's death. He ridiculed infinitesimals in his collections of mathematical curiosities, the *Apiaria universae philoso-phiae* of 1642 and the *Aerarium philosophiae mathematicae* of 1648.

Thomas Hobbes (1588–1679): The author of *Leviathan* and advocate of an absolutist authoritarian state, Hobbes considered himself a mathematician as well. He believed that his philosophy was founded on mathematical prin-ciples, and that it was therefore as certain as a geometrical demonstration. The decrees of the Leviathan, he believed, will be as incontestable as geo-metrical proofs.

André Tacquet (1612–60): Leading Jesuit mathematician, also charged with discrediting infinitesimals. He denounced infinitesimal mathematics in *Cyl-indricorum et annularium*, published in 1651, but accepted their limited use as heuristic devices. He was subsequently directed by his superiors to refrain from publishing original work, and to focus exclusively on writing textbooks. He did so.

JESUITS

Ignatius of Loyola (1491–1556): A Spanish nobleman and soldier from the Basque region, who experienced a religious awakening after being wounded in the Battle of Pamplona in 1521. With ten devoted followers he founded the Society of Jesus, which was officially recognized by Pope Paul III in 1540. Under his leadership the Jesuits became the most dynamic order of the Church and the most effective in battling the Reformation. By the time of Ignatius's death, the Society had grown to one thousand members and several dozen schools and colleges, and was still expanding rapidly.

Benito Pereira (1536–1610): Clavius's nemesis at the Collegio Romano, who insisted that mathematics did not qualify as a science. He was also the first Jesuit to directly condemn infinitesimals, although the context was not math-ematics, but rather a commentary on Aristotle.

Claudio Acquaviva (1543–1615): Superior general of the Jesuits from 1581 to 1615, Acquaviva established the office of the Revisors General, and sup-ported the early campaign against infinitesimals.

Mutio Vitelleschi (1563–1645): Superior general of the Jesuits from 1615 to 1645, during their period of eclipse (1623–31) and their return to power in Rome. He presided over the launch of the final campaign against infinitesi-mals, and wrote to the provinces to forbid the doctrine.

Jacob Bidermann (1578–1639): Leader of the Revisors General in 1632, when the Jesuits renewed their assault on infinitesimals.

Vincenzo Carafa (1585–1649): Superior general of the Jesuits from 1646 to 1649. Enforced the ban on infinitesimals, and humiliated Pallavicino by compelling him to retract his views. He wrote to his underlings to remain vigilant against infinitesimals, and began the process to add infinitesimals to a list of permanently banned doctrines.

Rodrigo de Arriaga (1592–1667): Leading Jesuit philosopher. In 1632 he published his *Cursus philosophicus*, which surprisingly concludes that infinitesimals are plausible. By the time the book was published, however, the Jesuits were back in power in Rome and determined to quash the support for infinitesimals. Superior General Carafa declared that there would be no more Arriagas.

Pietro Sforza Pallavicino (1607–67): A marchese (marquis) by birth, and an ardent Galilean in his youth, Pallavicino was exiled from Rome following Galileo's fall. He returned and, sensing which way the wind was blowing, became a Jesuit, and ultimately a cardinal. Still attached to Galilean views, he taught at the Collegio Romano that infinitesimals were plausible. In 1649 he was denounced in a letter by the superior general, and forced to publicly retract his views.

The Royal Society

Sir Francis Bacon (1561–1626): English jurist, philosopher, statesman, and Lord Chancellor to James I from 1618 to 1621. Although not a scientist himself, Bacon is considered one of the leading figures in the scientific revolution for his advocacy of empiricism in the study of nature. In a series of influential essays and books, Bacon argued that the proper way to investigate nature was through systematic observation and experimentation, rather than through a priori reasoning or mathematics. Long after his death, Bacon became the unofficial patron saint of the Royal Society, which championed his empiricist approach.

Henry Oldenburg (1619–77): German by birth, Oldenburg settled in London in the 1650s and became a key figure in intellectual and scientific circles, famous for his wide network of correspondents. Along with Robert Boyle, John Wallis, and others he was a leading founder of the Royal Society and served as its first secretary. In this capacity he guided the Society through its difficult early years, and established it as the leading scientific academy in Europe, known for its empiricist bent.

Robert Boyle (1627–91): A founder of modern chemistry, Boyle was the most distinguished and admired scientist among the early members of the Royal Society. Boyle advocated a humble empiricism as the correct approach to the study of nature, believing this would be to the benefit of both religion and the state.

Thomas Sprat (1635–1713): The leading propagandist of the early Royal Society. In 1667 he published the *History of the Royal Society of London*, which laid out the Society's scientific principles as well as its political goals. Sprat argued that the experimental study of nature not only increases human knowledge, but also promotes civic and religious harmony. In 1665 Sprat was the author of a sharp satirical response to Sorbière's account of his visit to England.

Rulers

Charles V (1500–58): Holy Roman Emperor from 1519 and king of Spain (as Charles I) from 1516 until his abdication in 1556. With domains stretching from eastern Europe to Peru, he was the titular ruler of one of the greatest empires in history, though his hold on his territories was often tenuous. Viewing himself as the sword of the Church, he confronted Luther in 1521 at the Diet (general assembly) of Worms, and issued an edict outlawing Luther and his teachings. He spent the rest of his reign trying unsuccessfully to eradicate Protestantism from his lands.

Gustavus Adolphus (1594–1632): King of Sweden from 1611, widely considered one of the greatest military innovators of all time. In June 1630 he landed with his army in northern Germany in support of the hard-pressed Protestant princes in the Thirty Years' War. Over the two years that followed, he proceeded to defeat the Catholic armies of the Holy Roman Empire in a string of battles, dramatically changing the balance of power in Europe. The Swedish threat also changed the political calculus in Rome, ending a period of Galilean ascendancy and handing power back to the Jesuits. Gustavus died in the Battle of Lützen while leading a cavalry charge against imperial forces.

Oliver Cromwell (1599–1658): One of the leading commanders of the Parliamentary New Model Army during the English Civil War, as well as leader of the Puritan Independents (against the Presbyterians). He became lord protector of England, Scotland, and Ireland in 1653. Some believed that Hobbes's *Leviathan* was written in support of his authoritarian rule.

Charles I (1600–49): King of England from 1625, his reign was marked by increasing confrontation with Parliament, and ultimately civil war. Following the example of the French kings, Charles tried to establish an absolute monarchy in England but encountered fierce resistance from Parliament, which controlled state revenues. His ill-fated attempt at personal rule led to the crisis of 1640 and the outbreak of civil war between Parliament and the king. Defeated in battle, Charles was captured by Parliamentary forces and executed in 1649.

Charles II (1630–85): The son of Charles I, he grew up in the court in exile, where he was tutored for a time by Hobbes. In 1660 he was recalled to England and reinstated as king by a coalition of former Parliamentarians and Royalists who were concerned about the rise of religious and social radicals. Careful to avoid the fate of his father, he ruled cautiously alongside Parliament. In 1662 he granted a royal charter to a group of natural philosophers who believed that the study of nature held the key to social and political peace. The group became the Royal Society of London.

POPES

Leo X (Pope from 1513 to 1521): A member of the Florentine Medici family, he was a cultured and learned man and a great patron of Renaissance art. But his slow and hesitant response to Luther's challenge helped turn a local problem in Germany into an existential crisis for the Church.

Paul III (Pope from 1534 to 1549): After ascending to the Papal throne at the height of the Reformation, when the Protestant tide was sweeping all before it, he set in motion the counteroffensive that would restore the fortunes of the Catholic Church. In 1540 he granted Ignatius of Loyola's request to found a new order called the Society of Jesus, which was destined to play a key role in the Counter-Reformation. In 1545 he convened the Council of Trent, which set the fundamental doctrines of the Catholic Church that are upheld to this day.

Gregory XIII (Pope from 1572 to 1585): A friend and protector of the Society of Jesus, he granted the Jesuits the land and resources to build a permanent home for their leading university, the Collegio Romano. He convened a commission for the reform of the calendar, in which Clavius played a key role, and implemented its recommendations in 1582. The Gregorian calendar, almost universally in use today, is named after him.

Urban VIII (Pope from 1623 to 1644): A friend and protector of Galileo's before ascending to the Papacy (known then as Cardinal Maffeo Barberini), Urban continued his patronage of Galileo as Pope, leading to a golden "liberal age" in Rome. But in 1632, following the publication of Galileo's *Dialogue* on the Copernican system and some unfavorable political developments, Urban turned on Galileo, leading to the latter's trial and banishment. The Jesuits returned to favor in Rome, and were given a free hand to suppress infinitesimals.

Clement IX (Pope from 1667 to 1669): As part of a short and undistinguished Papacy, Clement IX was responsible for suppressing the order of the Jesuats, of which two leading infinitesimalist mathematicians—Bonaventura Cavalieri and Stefano degli Angeli—were members.

OTHER REFORMERS, REVOLUTIONARIES, AND COURTIERS

Martin Luther (1483–1546): Originally an Augustinian monk and professor of theology at the University of Wittenberg, Luther launched the Reformation in 1517 by posting his Ninety-Five Theses on the door of the city's Castle Church. By 1621 he had been excommunicated by the Pope and banned by the emperor, but the spread of Protestantism proved irreversible. Other religious reformers soon followed Luther's lead, establishing their own brands of Protestantism.

Charles Cavendish (1594–1654): Respected mathematician and a member of one of the great aristocratic clans of England, known in the seventeenth century as patrons and practitioners of the arts and sciences. His brother William (1592–1676), the Duke of Newcastle, maintained a laboratory on the grounds of his estate, and William's wife, Margaret (1623–73), was an acclaimed poet and essayist. The Cavendishes turned their manors of Chatsworth and Welbeck Abbey into thriving intellectual centers, and were lifelong patrons of Hobbes.

Gerrard Winstanley (1609–76): Leader of the Diggers, who began digging up the land in St. Georges Hill, Surrey, in 1649. Winstanley and his followers believed that land was common property and that all men had the right to cultivate it. Their activities alarmed the local property owners, who managed to dislodge them through legal actions and violent attacks. Fear of the Diggers and other radical groups pushed the propertied classes to overcome their differences, leading to the restoration of the monarchy in 1660.

Samuel Sorbière (1615–70): French courtier, physician, and man of letters, as well as a friend and admirer of Thomas Hobbes. From 1663 to 1664 Sorbière visited England, spending much of the time as a guest of the Royal Society. His later account of the visit deeply offended his former hosts, especially his glorification of Hobbes and his ridicule of Wallis. This elicited a strong rebuttal from Thomas Sprat, and also ended Sorbière's career at the French court.

TIME LINE

Sixth century BCE: Pythagoras and his followers declare that "all is number," meaning that everything in the world can be described by whole numbers or their ratio.

Fifth century BCE: Democritus of Abdera uses infinitesimals to calculate the volume of cones and cylinders.

Fifth century BCE: The Pythagorean Hippasus of Metapontum discovers incommensurability (i.e., irrational numbers). It follows that different magnitudes are not composed of distinct tiny atoms, or infinitesimals. After his discovery, Hippasus is mysteriously lost at sea, possibly drowned at the hands of his Pythagorean brethren.

Fifth century BCE: Zeno of Elea proposes several paradoxes showing that infinitesimals lead to logical contradictions. Thereafter infinitesimals are shunned by ancient mathematicians.

300 BCE: Euclid publishes his highly influential treatise on geometry, *The Elements*, which carefully avoids infinitesimals. It serves as a model for the style and practice of mathematics for nearly two thousand years.

Ca. 250 BCE: Archimedes of Syracuse bucks the trend and experiments with infinitesimals. He arrives at remarkable new results regarding the areas and volumes enclosed by geometrical figures.

1517: Martin Luther launches the Reformation by nailing a copy of his Ninety-Five Theses to the door of the Castle Church in Wittenberg. The ensuing struggles between Catholics and Protestants continue for two centuries.

1540: Ignatius of Loyola founds the Society of Jesus, popularly known as the Jesuits, dedicated to reviving Catholic doctrines and restoring Church authority.

1544: A Latin translation of the works of Archimedes is published in Basel, making his study of infinitesimals widely available to scholars for the first time.

1560: Christopher Clavius begins teaching at the Jesuit Collegio Romano. He founds the Jesuit mathematical tradition on the bedrock of Euclidean geometry.

Late sixteenth to early seventeenth century: Revival of interest in infinitesimals among European mathematicians.

1601–15: The Jesuit "Revisors General," responsible for ruling on doctrines, produce a string of denunciations of infinitesimals.

1616: The Jesuits clash with Galileo over his advocacy of Copernicanism, but also for his use of infinitesimals. Galileo tones down his rhetoric, but bides his time for a chance to reopen the debates.

1616: The mathematician Luca Valerio sides with the Jesuits against his friend Galileo. He dies in disgrace soon after.

1618: Outbreak of the Thirty Years' War, pitting Catholics against Protestants.

1623: Galileo's friend Maffeo Barberini becomes Pope Urban VIII and sides openly with Galileo and his followers.

1623–31: A golden "liberal age" in Rome. Galileans ascendant.

1625–27: The Jesuit mathematician Gregory St. Vincent is prohibited by his superiors from publishing a work considered too close to infinitesimals.

1628: Thomas Hobbes encounters a geometrical proof for the first time while on a European tour.

1629: Bonaventura Cavalieri is appointed professor of mathematics at the University of Bologna.

1630s: Evangelista Torricelli develops his infinitesimal methods, but publishes nothing.

1631: The Protestant King Gustavus Adolphus of Sweden defeats the armies of the Holy Roman Emperor in the Battle of Breitenfeld during the Thirty Years' War. His victory alters the European balance of power.

1631: Under pressure from traditionalists, Urban renounces his liberal policies and restores the Jesuits to favor. End of Galilean ascendancy.

1632: The Jesuit Revisors General issue the most comprehensive condemnation of infinitesimals to date. Similar decrees follow in subsequent years.

1632: The Jesuit general superior, Mutio Vitelleschi, writes to the provinces to denounce infinitesimals.

1632–33: Galileo is charged with heresy, tried by the Inquisition, and condemned to spend the rest of his life under house arrest, which he does in his villa in Arcetri, outside Florence.

1635: Cavalieri publishes *Geometria indivisibilibus*, which becomes the standard work on infinitesimals across Europe.

1637: Galileo's *Dialogues Concerning Two New Sciences* is published in Leiden, in the Netherlands. The book discusses infinitesimals at length and praises Cavalieri as a "new Archimedes."

1640–60: The Interregnum. The civil war between King Charles I and Parliament leads to the execution of the king in 1649 and the establishment of a military dictatorship under Cromwell.

1640: Hobbes, a Royalist, flees to Paris and joins Charles I's court in exile, where he serves as mathematical tutor to the prince of Wales, the future Charles II.

1641: The Jesuit mathematician Paul Guldin publishes *De centro gravitatis*, which contains an attack on Cavalieri and a systematic critique of his method.

1642: Torricelli is appointed Galileo's successor at the Medici court and professor of mathematics at the Florentine Academy.

1642: Hobbes publishes his first philosophical work, *De cive*, in which he argues that only an absolute monarchy can save human society from chaos and civil war.

1644: Torricelli publishes his most important work on infinitesimals, the *Opera geometrica*.

1644: John Wallis is appointed secretary to the Westminster Assembly of Divines.

1645: Wallis joins with other science enthusiasts to conduct and discuss scientific experiments. The group, known as the "Invisible College," meets regularly for years.

1647: Cavalieri responds to Guldin in his last work, *Exercitationes geometricae sex*. He dies shortly thereafter.

1647: Death of Torricelli.

1648: The Peace of Westphalia ends the Thirty Years' War.

1648: The Jesuit mathematician Mario Bettini denounces infinitesimals in his book *Aerarium philosophiae mathematicae*.

1648: Pietro Sforza Pallavicino, Jesuit, nobleman, and future cardinal, is forced to publicly retract his advocacy of infinitesimals.

1649: Charles I of England is executed.

1649: Wallis appointed Savilian Professor of Mathematics at Oxford.

1649: The Jesuit superior general, Vincentio Carafa, writes to the provinces to denounce infinitesimals.

1651: The Jesuit mathematician André Tacquet, in *Cylindricorum et annularium libri IV*, declares that infinitesimals must be destroyed, or else mathematics will be destroyed.

1651: Hobbes publishes *Leviathan*, in which he advocates a totalitarian state. He grounds his reasoning in geometry.

1651: The Jesuits publish a list of permanently banned doctrines that includes infinitesimals.

1652: Hobbes falls out with the court in Paris and returns to England.

1655: Wallis publishes *De sectionibus conicis*.

1655: Hobbes publishes *De corpore*, which includes "proofs" of ancient unresolved problems such as squaring the circle.

1655: Wallis publishes *Elenchus geometriae Hobbianae*, in which he ridicules Hobbes and points out his mathematical mistakes.

1656: Wallis publishes *Arithmetica infinitorum*.

1656: Hobbes responds with *Six Lessons to the Professors of Mathematics*, wherein he retaliates by attacking Wallis's use of infinitesimals, which he considers nonsensical and conducive to error, not truth.

1657–79: Hobbes and Wallis criticize, ridicule, and hurl abuse at each other in dozens of books, pamphlets, and essays.

1658–68: Stefano degli Angeli, professor of mathematics at the University of Padua, publishes eight works on infinitesimals, all of them openly ridiculing the Jesuit critics of infinitesimal mathematics.

1660: Charles II is restored to the English throne.

1662: The "Invisible College" receives a charter from Charles II and becomes the Royal Society of London.

1665: The young Isaac Newton experiments with infinitesimals and develops a technique that will become known as the calculus.

1668: The monastic order of the Jesuats, home to Cavalieri and Angeli, suppressed by papal decree.

1675: Gottfried Wilhelm Leibniz develops his own version of the calculus.

1679: Hobbes dies, mathematically discredited and politically isolated.

1684: Leibniz publishes the first scholarly paper on the calculus, in the journal *Acta Eruditorum*.

1687: Newton publishes the *Principia mathematica*, revolutionizing physics and establishing the first modern theory of the solar system. The work is based on the infinitesimal calculus and contains Newton's first exposition of his method.

1703: Wallis dies, lauded as a leading mathematician, a forerunner of the calculus, and a founder of the Royal Society.

NOTES

Introduction

3 *French courtier Samuel Sorbière*: On Samuel Sorbière's visit to England, see his account in *A Voyage to England Containing Many Things Relating to the State of Learning, Religion, and Other Curiosities of That Kingdom* (London: J. Woodward, 1709), first published in Paris in 1664 as *Relation d'une voyage en Angleterre*. For the English reaction to Sorbière's account, see Thomas Sprat, *Observations on M. de Sorbière's Voyage into England* (London: John Martyn and James Allestry, 1665). A short biography of Sorbière is available in Alexander Chalmers, *General Biographical Dictionary* (London: J. Nichols and Son, 1812–17), 28:223. For a recent account of Sorbière's career and especially his visit to England, see Lisa T. Sarasohn, "Who Was Then the Gentleman? Samuel Sorbière, Thomas Hobbes, and the Royal Society," *History of Science* 42 (2004): 211–32.

3 *By his own testimony he was a "trumpeter"*: Sorbière's description of himself as a "trumpeter" and not a "soldier" is found in his dedication to King Louis XIV in *A Voyage to England*.

4 *"Hierarchy inspires People with Respect"*: Sorbière, *A Voyage to England*, pp. 23–24.

5 *"the Royal Society be not some way or other blasted"*: Ibid., p. 47.

5 *Sorbière insulted the Society's patron*: Quoted in Sarasohn, "Who Was Then the Gentleman?," p. 223.

5 *"noxious in conversation"*: Sorbière, *A Voyage to England*, p. 41.

5 *Hobbes, he wrote, was a courtly and "gallant" man*: For Sorbière's praise of Hobbes, see see ibid., pp. 40–41. Sprat's response is quoted in Sarasohn, "Who Was Then the Gentleman?," p. 225.

7 *"the indivisible Line of the Mathematicians"*: Sorbière, *A Voyage to England*, p. 93.

1. The Children of Ignatius

31 *Machiavelli's model of a cunning and brutal prince*: Machiavelli's advice to the ideal ruler is contained in *The Prince*, first published in Italian in 1532.

34 *"a soldier of God"*: For the founding of the Society of Jesus, see William V. Bangert, *A History of the Society of Jesus* (St. Louis, MO: Institute of Jesuit Sources, 1972). The quote is on page 21.

35 *the Society's growth*: Ibid., p. 98; R. Po-Chia Hsia, *The World of Catholic Renewal* (Cambridge: Cambridge University Press, 1998), p. 32.

36 *"a nursery of great men"*: The story of Montaigne's visit to the Collegio Romano is told in Bangert, *History of the Society of Jesus*, p. 56.

39 *their activities in France*: The Jesuits' troubles in France are discussed in Bangert, *History of the Society of Jesus*, esp. pp. 120–21.

40 *"to be obedient to the true Spouse of Christ"*: Ignatius of Loyola, "The First Rule," in "Rules for Thinking, Judging, and Feeling with the Church," in "The Spiritual Exercises," in *Spiritual Exercises and Selected Works* (Mahwah, NJ: Paulist Press, 1991), p. 111.

40 *"What I see as white"*: Loyola, "The Thirteenth Rule," in "The Spiritual Exercises," p. 213.

40 *Big Brother*: George Orwell, *1984*, part 3, chap. 2.

40 *"all authority is derived from God"*: An excellent discussion of the Jesuit ideal of obedience can be found in Steven Harris, "Jesuit Ideology and Jesuit Science," Ph.D. dissertation, University of Wisconsin–Madison, 1988, pp. 54–57, and also in Bangert, *History of the Society of Jesus*, p. 42.

41 *Neatness, cleanliness, and order*: The importance of neatness for Jesuits is discussed in Hermann Stoeckius, *Untersuchungen zur Geschichte der Noviziates in der Gesellschaft Jesu* (Bonn: P. Rost & Co., 1918). Quoted in Harris, "Jesuit Ideology," p. 83.

41 *"If the heretics should see"*: Favre's letter is quoted in Bangert, *History of the Society of Jesus*, p. 75.

46 *"Talis quus sis"*: The quote is from Francis Bacon, *The Advancement of Learning*, Book 1, III.3. Bacon is quoting from Plutarch's life of the Spartan king Agesilaus.

47 *the Jesuit intervention proved decisive*: On the Jesuits' work in Germany, Belgium, and Poland, see Hsia, *The World of Catholic Renewal*, chap. 4, "The Church Militant."

48 *"Your holy order"*: Pope Gregory XIII's address to the Jesuits can be found in Bangert, *History of the Society of Jesus*, p. 97.

48 *Rubens was a devout Catholic*: On Rubens and the Jesuits, see Hsia, *The World of Catholic Renewal*, pp. 128, 154.

51 *absolute authority*: On the Jesuits' ideas of authority derived from abso-

lute truth and expressed in the Church hierarchy, see Rivka Feldhay, "Authority, Political Theology, and the Politics of Knowledge in the Transition from Medieval to Early Modern Catholicism," *Social Research* 73, no. 4 (2006): 1065–92.

2. Mathematical Order

52 *"the parts of mathematics that a theologian should know"*: Ignatius's advice on the teaching of mathematics is quoted in Giuseppe Cosentino, "Mathematics in the Jesuit Ratio Studiorum," in Frederick J. Homann, SJ, ed. and trans., *Church Culture and Curriculum* (Philadelphia: St. Joseph University Press, 1999), p. 55.

52–53 *"to keep what remains, and restore what was lost"*: Polanco's letter is quoted in Cosentino, "Mathematics in the Jesuit Ratio Studiorum," p. 57.

53 *"a sort of hook with which we fish for souls"*: Nadal's pronouncement is quoted in M. Feingold, "Jesuits: Savants," in M. Feingold, ed., *Jesuit Science and the Republic of Letters* (Cambridge, MA: MIT Press, 2003), p. 6.

53 *Ignatius regarded him as well-nigh infallible*: On Ignatius and the setting of the curriculum at Jesuit colleges, see Cosentino, "Mathematics in the Jesuit Ratio Studiorum," p. 54.

55 *"anyone suspect us of trying to create something new"*: Pereira's and Acquaviva's views on innovation are quoted in Feingold, "Jesuits: Savants," p. 18.

55 Legem impone subactis: The impress of the Parthenic Academy is discussed in Ugo Baldini, *Legem impone subactis: Studi su filosofia e scienza dei Gesuiti in Italia 1540–1632* (Rome: Bulzoni, 1992), pp. 19–20.

56 *On April 12 he was received as a novice into the Society of Jesus by Ignatius of Loyola himself*: On Clavius's early years in the Society of Jesus, see James M. Lattis, *Between Copernicus and Galileo: Christoph Clavius and the Collapse of Ptolemaic Astronomy* (Chicago: University of Chicago Press, 1994), chap. 1.

57 *Clavius was self-taught*: On Clavius's mathematical education, see Lattis, *Between Copernicus and Galileo*, pp. 16–18.

57 *Even years later he was still fighting*: On Clavius's fight for recognition for mathematics professors in the Society, see A. C. Crombie, "Mathematics and Platonism in the Sixteenth-Century Italian Universities and in the Jesuit Educational Policy," in Y. Maeyama and W. G. Saltzer, eds., *Prismata* (Wiesbaden: Franz Steiner Verlag, 1977), pp. 63–94, esp. p. 65.

57 *his status in the rigid hierarchy of the order*: On Clavius's career at the Collegio Romano, see Cosentino, "Math in the Ratio Studiorum"; Crombie,

"Mathematics and Platonism," pp. 64–68; and Lattis, *Between Copernicus and Galileo*, chap. 1.

59 *"intolerable to all the wise"*: Quoted in J. D. North, "The Western Calendar: 'Intolerabilis, Horribilis, et Derisibilis,'" in G. V. Coyne, M. A. Hoskin, and O. Pedersen, eds., *Gregorian Reform of the Calendar: Proceedings of the Vatican Conference to Commemorate its 400th Anniversary, 1582–1982* (The Vatican: Pontificia Academia Scientarum, 1983), p. 75.

59 *it based its recommendations largely on Lilius's suggestions*: On Lilius and the calendar commission, see Ugo Baldini, "Christoph Clavius and the Scientific Scene in Rome," in Coyne, Hoskin, Pedersen, eds., *Gregorian Reform of the Calendar*, p. 137.

60 *Clavius would emerge as the public spokesman for the new system*: Clavius's report on the calendar commission was published as Christopher Clavius, *Romani calendarii a Gregorio XIII restituti explication*, first published in 1603. Clavius also wrote various pamphlets refuting critics such as Joseph Justus Scaliger, Michael Maestlin, and François Viète.

61 *"S. Stephen, John Baptist, & all the rest"*: Donne's satire was published as John Donne, *Ignatius His Conclave* (London: Nicholas Okes for Richard Moore, 1611). The passage is quoted in Lattis, *Between Copernicus and Galileo*, p. 8.

63 *Clavius believed that he knew what this secret was*: On Clavius's belief that it was mathematics that made the triumph of the calendar possible, see Romano Gatto, "Christoph Clavius' 'Ordo Servandus in Addiscendis Disciplinis Mathematicis' and the Teaching of Mathematics in Jesuit Colleges at the Beginning of the Modern Era," *Science and Education* 15 (2006): 235–36.

64 *"They demonstrate everything in which they see a dispute"*: See Clavius, "In disciplinas mathematicas prolegomena," in Christopher Clavius, ed., *Euclidis elementorum libri XV* (Rome: Bartholemaeum, 1589), p. 5. Clavius's *Euclid*, including the "Prolegomena," was first published in 1574.

64 *"The theorems of Euclid"*: See Clavius, "Prolegomena," p. 5, translation from Lattis, *Between Copernicus and Galileo*, p. 35.

64 *"the first place among the other sciences should be conceded to mathematics"*: Ibid.

65 *the angles in the same segment are all equal to one another*: These theorems appear in Euclid's *Elements* as propositions I.5, I.47, and III.21.

66 *Proposition 32*: See the instance in Dana Densmore, ed., *Euclid's Elements, the Thomas L. Heath Translation* (Santa Fe, NM: Green Lion Press, 2002), pp. 24–25.

67 *"Geometrical architect for all"*: The quote is from Antonio Possevino,

Biblioteca selecta (1591), quoted in Crombie, "Mathematics and Platonism," pp. 71–72.

68–69 *"mathematical disciplines are not proper sciences"*: Pereira continued: "to have science is to acquire knowledge of a thing through the cause on account of which the thing is; and science (scientia) is the effect of demonstration: but demonstration . . . must be established from those things that are 'per se,' and proper to that which is demonstrated, but the mathematician neither considers the essence of quantity, nor treats its affections as they flow from such essence, nor declares them by the proper causes on account of which they are in quantity, nor makes his demonstrations from proper and 'per se' but from common and accidental predicates." Benito Pereira, *De communibus omnium rerum naturalium principiis* (Rome: Franciscus Zanettus, 1576), I.12, p. 24. Quoted in Crombie, "Mathematics and Platonism," p. 67.

69 *If one seeks strong demonstrations, one must turn elsewhere*: For more on Pereira and the "Quaestio de certitudine mathematicarum," see Paolo Mancosu, "Aristotelian Logic and Euclidean Mathematics: Seventeenth-Century Developments in the Quaestio de Certitudine Mathematicarum," *Studies in History and Philosophy of Science* 23, no. 2 (1992): 241–65; Paolo Mancosu, *Philosophy of Mathematics and Mathematical ractice in the Seventeenth Century* (New York and Oxford: Oxford University Press, 1996); Gatto, "Christoph Clavius' 'Ordo Servandus,'" esp. pp. 239–42; and Lattis, *Between Copernicus and Galileo*, pp. 34–36.

69 *The subject of mathematics is matter itself*: The complete quote is "because the mathematical disciplines discuss things that are considered apart from any sensible matter—although they are themselves immersed in matter—it is evident that they hold a place intermediate between metaphysics and natural science. For the subject of metaphysics is separated from all matter, both in the thing and in reason; the subject of physics is in truth conjoined to sensible matter, both in the thing and in reason; whence, since the subject of the mathematical disciplines is considered free from all matter—although it [i.e., matter] is found in the thing itself—clearly it is established intermediate between the other two." Clavius, "Prolegomena," p. 5. Translation from Peter Dear, *Discipline and Experience* (Chicago: University of Chicago Press, 1995), p. 37.

70 *"It will contribute much"*: The passage is from Clavius, "Modus quo disciplinas mathematicas in scholis Societatis possent promoveri," quoted in Crombie, "Mathematics and Platonism," p. 66.

70 *Clavius also made positive suggestions*: The discussion and the quotes can be found in Clavius, "Modus quo disciplinas mathematicas," p. 65.

71 *"Ordo servandus"*: On Clavius's "Ordo servandus," see Gatto, "Christoph Clavius' 'Ordo Servandus,'" esp. pp. 243–46.

72 *Ratio studiorum*: On mathematics in the "Ratio studiorum," see Cosentino, "Math in the Ratio Studiorum," pp. 65–66.

72 *writing new textbooks*: For Clavius's textbooks and publications, see Gatto, "Christoph Clavius' 'Ordo Servandus,'" esp. pp. 243–44, and Ugo Baldini, "The Academy of Mathematics of the Collegio Romano from 1553 to 1612," in Feingold, ed., *Jesuit Science and the Republic of Letters*, esp. appendix C, pp. 74–75.

73 *a mathematics academy at the Collegio Romano*: On Clavius's academy at the Collegio Romano, see Baldini, "The Academy of Mathematics." On Acquaviva's decree, see Gatto, "Christoph Clavius' 'Ordo Servandus,'" p. 248.

74 *the Jesuits never deviated from their commitment to Euclidean geometry*: On the Jesuits' commitment to mathematics as both a key to understanding physical reality and as a model science to be emulated, see Rivka Feldhay, *Galileo and the Church: Political Inquisition or Critical Dialogue?* (Cambridge: Cambridge University Press, 1995), p. 222.

74 *the Dominicans could boast of no comparable accomplishment*: On the Jesuit-Dominican struggle for intellectual and theological supremacy, see Feldhay, *Galileo and the Church*.

75 *"some would rather be blamed by Clavius"*: Riccioli's comment is quoted in Eberhard Knobloch, "Sur la vie et l'oeuvre de Christophore Clavius (1538–1612)," *Revue d'histoire des sciences* 41, no. 3–4 (1988): 335.

75 *His many admirers outside the Society*: On Tycho Brahe's and the Archbishop of Cologne's opinions of Clavius, see ibid., pp. 335–36. On Commandino and Guidobaldo, see Mario Biagioli, "The Social Status of Italian Mathematicians, 1450–1600," *History of Science* 25 (1989): 63–64.

75 *"a German beast with a big belly"*: Clavius's detractors are quoted in Knobloch, "Sur la vie et l'oeuvre de Christophore Clavius," pp. 333–35.

77 *some new results in the theory of combinations*: On Clavius's innovations in this field, see ibid., pp. 343–51.

77 *a strict defender of the old orthodoxy*: On Clavius as a defender of orthodoxy, see Lattis, *Between Copernicus and Galileo*, as well as Knobloch, "Sur la vie et l'oeuvre de Christophore Clavius."

77 *he knew of Viète's groundbreaking work*: On the absence of Viète's analysis from Clavius's *Algebra*, see Baldini, "The Academy of Mathematics," p. 63.

77 *"The theorems of Euclid and the rest of the mathematicians"*: Clavius, "Prolegomena," p. 5, translation from Lattis, *Between Copernicus and Galileo*, p. 35.

78 *even suspect the Jesuits of innovating*: Acquaviva's warning is quoted in Feingold, "Jesuits: Savants," p. 18.

3. Mathematical Disorder

81 *the Jesuits celebrated Galileo*: The story of Galileo's astronomical discoveries and his triumphant visit to Rome and the Collegio Romano is summarized in Stillman Drake, ed. and trans., *Discoveries and Opinions of Galileo* (New York: Anchor Books, 1957).

81 *Discourse on Floating Bodies*: *Discourse* is discussed ibid., pp. 79–81.

81 *Letters on Sunspots*: The *Letters* are discussed and largely translated ibid., pp. 59–144.

83 *"Letter to the Grand Duchess Christina"*: Discussed and translated ibid., pp. 145–216.

86 *"I call the lines so drawn 'all the lines'"*: Cavalieri to Galileo, December 15, 1621. The letter can be found in Bonaventura Cavalieri, *Lettere a Galileo Galilei*, ed. Paolo Guidera (Verbania, Italy: Caribou, 2009), pp. 9–10.

87 *As early as 1604*: Galileo's early work on infinitesimals is cited in Festa, "La querrelle de l'atomisme," p. 1042; and Festa, "Quelques aspects de la controverse sur les indivisibles," p. 196.

87 *Galileo was still occupied with the paradoxes of the continuum*: See Festa, "La querrelle de l'atomisme," p. 1043.

87 *Dialogue on the Two Chief World Systems*: The standard English edition is Galileo, *Dialogue on the Two Chief World Systems*, trans. Stillman Drake (Berkeley: University of California Press, 1962), first published in Florence in 1632.

90 *the mathematical continuum was modeled on physical reality*: In addition to the discussion in the *Discourses*, Galileo presented a similar view of the continuum in his commentary on Antonio Rocco's *Filosofiche esercitazioni* in 1633. See François de Gandt, "Naissance et métamorphose d'une théorie mathématique: La géométrie des indivisibles en Italie," *Sciences et techniques en perspective*, vol. 9, *1984–85* (Nantes: Université de Nantes, 1985), p. 197.

91 *he had experimented extensively with indivisibles*: On Valerio's work on indivisibles, see Carl B. Boyer, *The History of the Calculus and Its Conceptual Development* (New York: Dover Publications, 1949), pp. 104–106; and the entry "Luca Valerio" in The MacTutor History of Mathematics archive mathematical biographies, http://www-history.mcs .st-andrews.ac.uk/Biographies/Valerio.html.

93 *It follows, Salviati concludes*: Salviati's discussion of the law of falling bodies can be found in Galileo, *Dialogues Concerning Two New Sciences*, ed. Henry Crew and Alfonso de Salvio (Buffalo, NY: Prometheus Books, 1991), pp. 173–74. In the standard Italian Edizione Nazionale, it is on pp. 208–209.

94 *"I am proud, and will always be"*: Quoted in Enrico Giusti, *Bonaventura*

Cavalieri and the Theory of Indivisibles (Bologna: Edizioni Cremonese, 1980), p. 3n9.

94 *"his singular inclinations and ability"*: Quoted ibid., p. 3n10.

95 *"I am now in my own country"*: Cavalieri to Galileo, July 18, 1621, quoted ibid., p. 6n18.

95 *"to the great wonder of everyone"*: The quote is from Cavalieri's biographer Girolamo Ghilini, quoted ibid., p. 7n19.

95 *he approached the Jesuit fathers*: Cavalieri to Galileo, August 7, 1626, quoted ibid., p. 9n26.

95 *"few scholars since Archimedes"*: Galileo to Cesare Marsili, March 10, 1629, quoted ibid., p. 11n30.

96 *"like cloths woven of parallel threads"*: Cavalieri's comparison of indivisibles to threads in a cloth and pages in a book, and his discussion of the metaphors, are found in his *Exercitationes geometricae sex* (Bologna: Iacob Monti, 1647), pp. 3–4.

97 *proposition 19*: Cavalieri, *Geometria indivisibilibus libri VI*, proposition 19, pp. 437–39. For discussions of this proof, see de Gandt, "Naissance et métamorphose d'une théorie mathématique," pp. 216–17; and Margaret E. Baron, *The Origins of the Infinitesimal Calculus* (New York: Dover Publications, 1969), pp. 131–32.

99 *Archimedes had used his own ingenious approach*: On Archimedes's calculation of the area enclosed in a spiral, see Baron, *The Origins*, pp. 43–44.

100 *Cavalieri's calculation of the area enclosed inside a spiral*: The calculation is presented as proposition 19 in Cavalieri, *Geometria indivisibilibus libri VI*, p. 238.

102 *"It can be said that with the escort of good geometry"*: Cavalieri to Galileo, June 21, 1639, in Galileo, *Opere*, vol. 18, p. 67, letter no. 3889. Quoted and translated in Amir Alexander, *Geometrical Landscapes* (Stanford, CA: Stanford University Press, 2002), p. 184.

102 *"all the lines"*: On the importance of the concepts of "all the lines" and "all the planes" in Cavalieri's work, see de Gandt, "Naissance et métamorphose d'une théorie mathématique," and François de Gandt, "Cavalieri's Indivisibles and Euclid's Canons," in Peter Barker and Roger Ariew, eds., *Revolution and Continuity: Essays in the History and Philosophy of Early Modern Science* (Washington, DC: Catholic University of America, 1991), pp. 157–82; and Giusti, *Bonaventura Cavalieri*.

104 *Cavalieri's younger contemporary*: The biographical information on Torricelli is from Egidio Festa, "Repères biographique et bibliographique," in François de Gandt, ed., *L'oeuvre de Torricelli: Science Galiléenne et nouvelle géométrie* (Nice: CNRS and Université de Nice, 1987), p. 8.

104 *"I was the first in Rome"*: Torricelli to Galileo, September 11, 1632, quoted ibid.

105 *"how the road you have opened"*: Castelli to Galileo, March 2, 1641, quoted ibid., p. 9.

106 *a long and fruitful correspondence with French scientists*: On Torricelli's correspondence with his French colleagues, see Armand Beaulieu, "Torricelli et Mersenne," in de Gandt, ed., *L'oeuvre de Torricelli*, pp. 39–51. On his connections with Italian Galileans, see Lanfranco Belloni, "Torricelli et son époque," in de Gandt, ed., *L'oeuvre de Torricelli*, pp. 29–38; on the barometer, see Festa, "Repères," pp. 15–18, and P. Souffrin, "Lettres sur la vie," in de Gandt, ed., *L'oeuvre de Torricelli*, pp. 225–30.

106 *Opera geometrica*: Torricelli's *Opera geometrica* can be found in volume 1 of Gino Loria and Giuseppe Vassura, eds., *Opere di Evangelista Torricelli* (Faenza: G. Montanari, 1919–44). An Italian translation is in Lanfranco Belloni, ed., *Opere scelte di Evangelista Torricelli* (Turin: Unione Tipografico-Editrice Torinese, 1975), pp. 53–483.

107 *"De dimensione parabolae"*: For discussions of "De dimensione parabolae," see François de Gandt, "Les indivisibles de Torricelli," in de Gandt, ed., *L'oeuvre de Torricelli*, pp. 152–53; and in de Gandt, "Naissance et métamorphose," pp. 218–19.

109 *that the ancients possessed a secret method*: Torricelli discusses this idea in the *Opera geometrica*, esp. in Loria and Vassura, eds., *Opere*, vol. 1, pp. 139–40. The Italian translation is in Belloni, ed., *Opere scelte*, p. 381.

110 *"the Royal Road through the mathematical thicket"*: See Loria and Vassura, eds., *Opere*, vol. 1, p. 173. Quoted in de Gandt, "Les indivisibles de Torricelli," p. 153.

110 *"We turn away from the immense ocean of Cavalieri's Geometria"*: Loria and Vassura, eds., *Opere*, vol. 1, p. 141.

111 *Torricelli's directness made his method far more intuitive*: See de Gandt, "Naissance et métamorphose," p. 219.

111 *three separate lists of paradoxes*: On Torricelli's lists of paradoxes, see de Gandt, "Les indivisibles de Torricelli," pp. 163–64.

112 *the simplest one captures the essential problem*: Torricelli's basic paradox is presented in a treatise entitled "De indivisibilium doctrina perperam usurpata," in Loria and Vassura, eds., *Opere*, vol. 1, part 2, p. 417.

113 *"that indivisibles are all equal to each other"*: Torricelli's discussion of unequal indivisibles can be found in Loria and Vassura, eds., *Opere*, vol. 1, part 2, p. 320. It is quoted in de Gandt, "Les indivisibles de Torricelli," p. 182.

114 *"semi-gnomons"*: The diagrams here are derived from de Gandt, "Les indivisibles de Torricelli," p. 187.

115 *to calculate the slope of the tangent*: For a discussion of Torricelli's method of tangents, see ibid., pp. 187–88, and idem, "Naissance et

métamorphose," pp. 226–29. De Gandt's exposition is based on Torricelli's *Opere*, ed. Loria and Vassura, vol. 1, part 2, pp. 322–33.

4. "Destroy or Be Destroyed": The War on the Infinitely Small

118 *"one of the best books ever written on mathematics"*: Quoted in H. Bosmans, "André Tacquet (S. J.) et son traité d'arithmétique théorique et pratique," *Isis* 9 (1927): 66–82.

119 *"either legitimate or geometrical"*: André Tacquet, *Cylindricorum et annularium libri IV* (Antwerp: Iacobum Mersium, 1651), pp. 23–24.

119 *geometry formed the core of Jesuit mathematical practice*: On the persistence of the Euclidean tradition among Jesuit mathematicians through the eighteenth century, see Bosmans, "André Tacquet," p. 77. Not coincidentally, some of the most popular textbooks on Euclidean geometry in that era were composed by Jesuits, including Honoré Fabri, *Synopsis geometrica* (Lyon: Antoine Molin, 1669); and Ignace-Gaston Pardies, *Elémens de géométrie* (Paris: Sébastien Maire-Cramoisy, 1671). Both textbooks were published repeatedly in the seventeenth and eighteenth centuries.

120 *the struggle between geometry and indivisibles*: Tacquet, *Cylindricorum et annularium*, pp. 23–24.

121 *the continuum is infinitely divisible*: Benito Pereira, *De communibus omnium rerum naturalium principiis* (Rome: Franciscus Zanettus, 1576). On Pereira's discussion of the composition of the continuum, see Paolo Rossi, "I punti di Zenone," *Nuncius* 13, no. 2 (1998): 392–94.

122 *"great confusion and perturbation"*: Quoted in Feingold, "Jesuits: Savants," p. 30.

122 *The first decree by the Revisors General*: The Revisors' condemnations of 1606 and 1608 can be found in the Jesuit archive ARSI (Archivum Romanum Societatis Iesu), manuscript FG656 A I, pp. 318–19.

123 *Luca Valerio of the Sapienza University*: The book was Luca Valerio, *De centro gravitatis solidorum libri tres* (Rome: B. Bonfadini, 1604). On Valerio's use of infinitesimal methods, see Carl B. Boyer, *History of the Calculus* (New York: Dover Publications, 1947), pp. 104ff.

123 *experimenting with indivisibles*: On Galileo's 1604 experimentation with indivisibles see Festa, "Quelques aspects de la controverse sur les indivisibles," p. 196.

123 *"the Archimedes of our age"*: Galileo, *Dialogues Concerning Two New Sciences*, p. 148.

125 *"the continuum is composed of indivisibles"*: The first condemnation of 1615, dated April 4, is in ARSI manuscript FG 656A II, p. 456. The second, dated November 19, is in manuscript FG 656A II, p. 462.

127 *Valerio had misread the signs*: On Valerio's rise and fall, see David Freed-
    berg, *The Eye of the Lynx: Galileo, His Friends, and the Beginnings of
    Modern Natural History*, (Chicago: University of Chicago Press, 2002),
    esp. pp. 132–34, as well as the online MacTutor biography of Valerio by
    J. J. O'Connor and E. F. Robertson at http://www.gap-system.org/~history
    /Biographies/Valerio.html. On his use of infinitesimals, see Boyer, *His-
    tory of the Calculus*, pp. 104–106.

128 *By relying on their hierarchical order*: On tensions between Jesuit intel-
    lectuals and the hierarchy's efforts to control their scholarly work, see
    Feingold, "Jesuits: Savants"; and Marcus Hellyer, "'Because the Author-
    ity of My Superiors Commands': Censorship, Physics, and the German
    Jesuits," *Early Science and Medicine* 1, no. 3 (1996): 319–54.

129 *he settled for a curt permission by the Jesuit provincial of Flanders*: In
    contrast, when Tacquet published his *Cylindricorum et annularium* four
    years later, also in Flanders, his license stated that his work had been
    read and approved by three mathematicians of the Society. St. Vincent's
    book carried no such endorsement.

129 *St. Vincent's experience typifies the Jesuit attitude toward indivisibles*: On
    Gregory St. Vincent's troubles, see Feingold, "Jesuits: Savants," pp.
    20–21; Herman Van Looey, "A Chronology and Historical Analysis of
    the Mathematical Manuscripts of Gregorius a Sancto Vincentio
    (1584–1667)," *Historia Mathematica* 11 (1984): 58; Bosmans, "André
    Tacquet," pp. 67–68; also Paul B. Bockstaele, "Four Letters from Gre-
    gorius a S. Vincentio to Christopher Grienberger," *Janus* 56 (1969):
    191–202.

130 *For the Jesuits, the choice could hardly have been worse*: On the changing
    political and cultural climate in Rome surrounding the election of Ur-
    ban VIII, see Pietro Redondi, *Galileo Heretic* (Princeton, NJ: Princeton
    University Press, 1987), esp. pp. 44–61 and 68–106.

131 *"bring down the pride of the Jesuits"*: This is from a letter by the Lincean
    Dr. Johannes Faber to Galileo from February 15, 1620. Quoted in Re-
    dondi, *Galileo Heretic*, p. 43.

131 *the new Pope was amused and full of admiration*: Cesarini to Galileo,
    October 28, 1623. Quoted in Redondi, *Galileo Heretic*, p. 49.

133 *The plot misfired badly*: On the Santarelli affair, see Bangert, *A History
    of the Society of Jesus*, pp. 200–201; and Redondi, *Galileo Heretic*, pp.
    104–105.

133 *Galileo even believed that he had been given implicit permission*: See Re-
    dondi, *Galileo Heretic*, p. 50.

134 *The young aristocrat was a rising star in Roman intellectual circles*: On
    Pallavicino's dissertation defense, see ibid., pp. 200–202. Father Grassi
    had been Galileo's opponent in a dispute over the nature of comets and
    the chief target of *The Assayer*. His condemnation of the orthodoxy of

Galileo's atomism was contained in his 1626 book *Ratio ponderum librae et simbellae*, published under the pseudonym Lothario Sarsi.

137 *Urban VIII had run out of options*: On the political crisis in Rome in 1631, and on Cardinal Borgia's attack on Urban VIII, see Redondi, *Galileo Heretic*, pp. 229–31.

138 *Galileo's friends were running for cover*: Not all made it safely. In April of 1632, Giovanni Ciampoli, the most prominent Lincean in the Curia and personal secretary to the Pope himself, was given the impressive-sounding title of "Governor of Montalti di Castro" and exiled from Rome to the Apennines. He would never return.

139 *Father Rodrigo de Arriaga of Prague*: On Rodrigo Arriaga, his *Cursus philosophicus*, and his views on the infinitely small, see Rossi, "I punti di Zenone," pp. 398–99; Hellyer, "'Because the Authority of My Superiors Commands,'" p. 339; Feingold, "Jesuits: Savants," p. 28; Redondi, *Galileo Heretic*, pp. 241–42; and John L. Heilbron, *Electricity in the 17th and 18th Centuries* (Berkeley: University of California Press, 1979), p. 107.

140 *Quite possibly he was influenced by his friend Gregory St. Vincent*: On Arriaga's friendship with St. Vincent, see Van Looey, "A Chronology and Historical Analysis of the Mathematical Manuscripts of Gregorius a Sancto Vincentio," p. 59.

140 *"The permanent continuum can be constituted"*: The Revisors' decree is preserved as manuscript FG 657, p. 183, in ARSI (Archivum Romanum Societatis Iesu), the archive of the Society of Jesus in Rome. It is also reproduced in Egidio Festa, "La querelle de l'atomisme," *La Recherche* 224 (September 1990): 1040; and quoted in French in Egidio Festa, "Quelques aspects de la controverse sur les indivisibles," in M. Bucciantini and M. Torrini, eds., *Geometria e atomismo nella scuola Galileana* (Florence: Leo S. Olschki, 1992), p. 198. Special thanks to Professor Carla Rita Palmerino of Radboud University Nijmegen, in the Netherlands, for making available to me her notes from the Jesuit archives.

141 *he found himself writing to Father Ignace Cappon*: General Mutio Vitelleschi to Ignace Cappon, 1633, quoted in Michael John Gorman, "A Matter of Faith? Christoph Scheiner, Jesuit Censorship, and the Trial of Galileo," *Perspectives on Science* 4, no. 3 (1996): pp. 297–98. Also quoted in Feingold, "Jesuits: Savants," p. 29.

142 *On February 3, 1640*: ARSI manuscript FG 657, p. 481.

142 *in January 1641*: Ibid., p. 381. Cited and discussed in Festa, "Quelques aspects," pp. 201–202.

142 *On May 12, 1643*: ARSI manuscript FG 657, p. 395.

143 *In 1649*: Ibid., p. 475.

144 *Arriaga's views on the continuum were unequivocally condemned*: On Arriaga and the publishing history of his *Cursus philosophicus*, see Hellyer, "'Because the Authority of My Superiors Commands,'" pp. 339–41.

145 *Pallavicino was no ordinary novice*: On Pietro Sforza Pallavicino and his career, see Redondi, *Galileo Heretic*, pp. 264–65; Hellyer, "'Because the Authority of My Superiors Commands,'" p. 339; Festa, "La querelle de l'atomisme," pp. 1045–46; Festa, "Quelques aspects," pp. 202–203; and Feingold, "Jesuits: Savants," p. 29.

145 *the marchese still considered himself a progressive thinker*: Redondi, *Galileo Heretic*, p. 265.

145 *Pallavicino frequently came under the Revisors' scrutiny*: On Pallavicino's conflicts with the Revisors and General Carafa, see Claudio Costantini, *Baliani e i Gesuiti* (Florence: Giunti, 1969), esp. pp. 98–101.

145 *Pallavicino forged ahead, lecturing on his unorthodox views*: Pallavicino hints at his troubles in Pietro Sforza Pallavicino, *Vindicationes Societatis Iesu* (Rome: Dominic Manephi, 1649), p. 225. Quoted and discussed in Festa, "Quelques aspects," pp. 202–203.

146 *"there are some in the Society who follow Zeno"*: Superior General Vincenzo Carafa to Nithard Biberus, March 3, 1649. In G. M. Pachtler, SJ, ed., *Ratio studiorum et institutiones scholasticae Societatis Jesu* (Osnabrück: Biblio-Verlag, 1968), 3:76, doc. no. 41.

147 *Ordinatio pro studiis superioribus*: For the text of the *Ordinatio*, see G. M. Pachtler, SJ, ed., *Ratio studiorum*, vol. 3 (Berlin: Hofman and Comp., 1890), pp. 77–98. The sixty-five banned "philosophical" propositions are on pages 90–94, and an additional list of twenty-five banned "theological" propositions is on pages 94–96. For a discussion of the *Ordinatio*, its origins, and its effects, see Hellyer, "'Because the Authority of My Superiors Commands,'" pp. 328–29. It is also mentioned in Feingold, "Jesuits: Savants," p. 29; and Carla Rita Palmerino, "Two Jesuit Responses to Galileo's Science of Motion: Honoré Fabri and Pierre le Cazre," in M. Feingold, ed., *The New Science and Jesuit Science: Seventeenth-Century Perspectives* (Dordrecht: Kluwer Academic Publishers, 2003), p. 187.

147 *"The succession continuum"*: The propositions are listed in Pachtler, ed., *Ratio studiorum*, p. 92.

5. The Battle of the Mathematicians

150 *"the three Jesuits, Guldin, Bettini, and Tacquet"*: Stefano degli Angeli, *De infinitis parabolis* (Venice: Ioannem La Nou, 1659), under "Lectori Benevolo."

151 *It was also crucial to prove them mathematically wrong*: On Guldin, Bettini, and Tacquet as the Society's agents sent to combat the method of indivisibles, see Redondi, *Galileo Heretic*, p. 291.

152 *Guldin was Clavius's follower*: For an excellent short biography of Guldin (and many other mathematicians), see the online MacTutor History of

Mathematics archive, hosted by the University of St. Andrews in Scotland, at http://www-history.mcs.st-and.ac.uk/. See also Guldin's biography authored by J. J. O'Connor and E. F. Robertson at http://www.gap-system.org/~history/Biographies/Guldin.html.

152 *He first suggests that Cavalieri's method is not in fact his own*: On Guldin's charge that Cavalieri derived his method from Kepler and Sover, see Giusti, *Bonaventura Cavalieri*, pp. 60–62; and Mancosu, *Philosophy of Mathematics and Mathematical Practice*, pp. 51–52.

153 *"no geometer will grant him"*: Paul Guldin, *De centro gravitatis*, book 4 (Vienna: Matthaeus Cosmerovius, 1641), p. 340.

153 *This then leads Guldin to his final point*: On Guldin's mathematical criticisms of Cavalieri and his method, see Giusti, *Bonaventura Cavalieri*, pp. 62–64; and Mancosu, *Philosophy of Mathematics*, pp. 50–55.

154 *"reasons that must be suppressed"*: Guldin, *De centro gravitatis*, book 2 (Vienna: Matthaeus Cosmerovius, 1639), p. 3. Quoted in Bonaventura Cavalieri, *Exercitationes geometricae sex* (Bologna: Iacob Monti, 1647), p. 180, and quoted and discussed in Festa, "Quelques aspects," p. 199.

155 *Initially he intended to respond in the form of a dialogue*: On Cavalieri's plans for a dialogue and Rocca's advice, see Giusti, *Bonaventura Cavalieri*, pp. 57–58.

155 *None of this, he argues, has any bearing on the method of indivisibles*: For Cavalieri's claim to be agnostic on the subject of the composition of the continuum, see Mancosu, *Philosophy of Mathematics*, p. 54.

156 *"relative infinity"*: See Mancosu, *Philosophy of Mathematics*, p. 54; and Giusti, *Bonaventura Cavalieri*, p. 64.

156 *"it is not necessary to describe actually"*: Cavalieri, *Exercitationes geometricae sex*, part 3, "In Guldinum," quoted in Giusti, *Bonaventura Cavalieri*, pp. 62–63.

157 *"the hand, the eye, or the intellect?"*: Guldin, *De centro gravitatis*, book 4, p. 344, quoted in Giusti, *Bonaventura Cavalieri*, p. 63.

157 *Mario Bettini, who inherited the mantle from Guldin*: On Bettini, his place among the Jesuits and his relationship with Christoph Grienberger, see Michael John Gorman, "Mathematics and Modesty in the Society of Jesus: The Problems of Christoph Grienberger," in Feingold ed., *The New Science and Jesuit Science*, pp. 4–7.

157 *the author of two very long and eclectic books*: Mario Bettini, *Apiaria universae philosophiae mathematicae* (Bologna: Io. Baptistae Ferronij, 1645); Mario Bettini, *Aerarium philosophiae mathematicae* (Bologna: Io. Baptistae Ferronij, 1648).

158 *"were the Jesuit Fathers not here"*: Cavalieri to Galileo, August 7, 1626, quoted in Giusti, *Bonaventura Cavalieri*, p. 9.

158 *The move was ultimately blocked by the city's senate*: See Giusti, *Bonaventura Cavalieri*, pp. 9–10n26.

159 *"infinity to infinity has no proportion"*: Guldin, *De centro gravitatis*, book 4, p. 341, quoted in Mancosu, *Philosophy of Mathematics*, p. 54.

159 *"'what separates the false coin from the true'"*: Bettini, *Aerarium*, vol. 3, book 5, p. 20. The quote is from Horace, *Epistles*, book 1.7, line 23: "Quid distent aera lupinis."

159 *"I respond to the counterfeit philosophizing"*: Quoted in Stefano degli Angeli, "Appendix pro indivisibilibus," in *Problemata geometrica sexaginta* (Venice: Ioannem la Nou, 1658), p. 295.

159 *Cylindricorum et annularium*: André Tacquet, *Cylindricorum at annularium libri IV* (Antwerp: Iacobus Meurisius, 1651).

161 *the general's response was surprisingly cool*: On Tacquet and General Nickel, see Bosmans, "André Tacquet," p. 72.

162 *"a noble geometer"*: Tacquet, *Cylindricorum*, pp. 23–24, quoted and discussed in Festa, "Quelques aspects," pp. 204–205.

162 *"nothing can be proven by anyone"*: Tacquet, *Cylindricorum*, p. 23.

162 *"I will always doubt its truth"*: Ibid., p. 24, quoted and discussed in Bosmans, "André Tacquet," p. 72.

164 *Cavalieri's student at Bologna*: On Aviso and Mengoli, see Giusti, *Bonaventura Cavalieri*, pp. 49–50, as well as the Cavalieri and Mengoli entries in Charles Gillispie, ed., *Dictionary of Scientific Biography* (New York: Scribner, 1981–90).

164 *Vincenzo Viviani (1622–1703)*: On Viviani, see Giusti, *Bonaventura Cavalieri*, p. 51, at J. J. O'Connor and E. F. Robertson, "Vincenzo Viviani," MacTutor online biography at http://www-groups.dcs.st-and.ac.uk/history/Biographies/Viviani.html.

165 *Antonio Nardi*: On Nardi, see Giusti, *Bonaventura Cavalieri*, p. 51, as well as Belloni, "Torricelli et son époque," pp. 29–38.

167 *His first broadside*: Angeli, "Appendix pro indivisibilibus."

168 *"can be called The Bee"*: Angeli's discussion of Bettini as a busy but unlucky bee can be found in his *Problemata*, pp. 293–95.

169 *Everyone, he responds, except the Jesuits*: Angeli's polemic against Tacquet and his fellow Jesuit mathematicians is included in his preface to the reader in Stefano degli Angeli, "Lectori Benevolo," in *De infinitis parabolis*.

171 *"no advantage or utility to the Christian people"*: On the papal brief of 1668 suppressing the three orders, see Sydney F. Smith, SJ, Joseph A. Munitiz, SJ, eds., *The Suppression of the Society of Jesus* (Eastbourne, UK: Antony Rowe Ltd., 2004), pp. 291–92. First published as a series of articles by Sydney Smith in *The Month* between February 1902 and August 1903.

172 *the "Aquavitae Brothers"*: See William Eamon, "The Aquavitae Brothers," in http://williameamon.com/?p=552; and T. Kennedy, "Blessed John Colombini," in *The Catholic Encyclopedia* (New York: Robert Appleton Company, 1910).

174 *he had previously published no fewer than nine books*: Angeli's books
were *Problemata geometrica sexaginta* (1658); *De infinitis parabolis* (1659);
Miscellaneum hyperbolicum et parabolicum (1659); *Miscellaneum geome-
tricum* (1660); *De infinitorum spiralium spatiorum mensura* (1660); *De infinitorum cochlearum mensuris* (1661); *De superficie ungulae* (1661);
Accessionis ad stereometriam et mecanicam (1662); and *De infinitis spi-
ralibus inversis* (1667). See Giusti, *Bonaventura Cavalieri*, p. 50n39.

176 *Galileo was a brilliant public advocate for the freedom to philosophize*:
The quote is from Galileo Galilei, "Third Letter on Sunspots," in Drake,
ed., *Discoveries and Opinions of Galileo*, p. 134. A translation of the
"Letter to the Grand Duchess Christina" can be found in the same volume, pp. 173–216.

178 *Italy had been home to perhaps the liveliest mathematical community in
Europe*: On the early modern mathematical tradition in Italy, see Mario
Biagioli, "The Social Status of Italian Mathematicians, 1450–1600," *His-
tory of Science* 27, no. 1 (1989): 41–95.

6. The Coming of Leviathan

183 *yet they went on digging*: The story of the Diggers is told in Christopher
Hill, *The World Turned Upside Down* (Harmondsworth: Penguin
Books, 1975), chap. 7, "Levellers and True Levellers." Quotes are from
p. 110.

184 *the Diggers soon followed up with a pamphlet*: The pamphlet was called
The True Levellers Standard Advanced, printed in 1649.

191 *many other groups, and unnumbered individuals, emerged to take their
place*: For a detailed account of the radical sects of the English Revolution, see Hill, *The World Turned Upside Down*.

193 *"a giddy hot-headed, bloody multitude"*: The comment is by the Reverend
Henry Newcombe, quoted in Christopher Hill, *The Century of Revolution, 1603–1714* (New York: W. W. Norton and Company, 1982), p. 121.

193 *"the gentry and citizens throughout England"*: Pepys is quoted in Hill,
Century of Revolution, p. 121.

194 *"both Me, and Fear"*: Quoted in Samuel I. Mintz, *The Hunting of Levia-
than: Seventeenth-Century Reactions to the Materialism and Moral
Philosophy of Thomas Hobbes* (Cambridge, UK: Cambridge University
Press, 1970), p. 1.

195 *"a little learning went a great way with him"*: John Aubrey's biography of
Hobbes can be found as "Thomas Hobbes," in Andrew Clark, ed., *"Brief
Lives," Chiefly of Contemporaries, Set down by John Aubrey, between the
Years 1669 and 1696* (Oxford: Clarendon Press, 1898), pp. 321–403. The
quote is from p. 391.

195 *a laudable record of sobriety*: Aubrey's report on Hobbes's drinking can be found in his biography "Thomas Hobbes," p. 350.

195 *"prove things after my owne taste"*: This is quoted in Mintz, *Hunting of Leviathan*, p. 2.

198 *"the one who has opened to us the gate"*: Ibid., pp. 8–9.

199 *Thomas Harriot*: On Harriot, see Alexander, *Geometrical Landscapes*.

200 *tutor to the Prince of Wales*: On Hobbes's appointment as royal tutor and the opposition to it, see Mintz, *Hunting of Leviathan*, p. 12.

201 *Leviathan*: Thomas Hobbes, *Leviathan, or the Matter, Forme, and Powers of a Commonwealth Ecclesiastical and Civil* (London: Andrew Crooke, 1651).

202 *"he cannot assure the power and means to live well"*: See Hobbes, *Leviathan*, 11:2.

203 *"solitary, poor, nasty, brutish, and short"*: This famous quote appears ibid., 13:9.

203 *"war of every man against every man"*: Ibid., 13:13.

204 *"the savage people in many places of America"*: For Hobbes's view of native Americans as living in the state of nature, see ibid., 13:11.

204 *"more than consent, or concord"*: Ibid., 17:13.

204 *"This is the generation of that great LEVIATHAN"*: Ibid.

204 *"one person, of whose acts a great multitude"*: Ibid.

206 *"is but an artificial man"*: Introduction ibid., p. 1.

207 *blaming them directly for the onset of the civil war*: Reflecting on the role of clergymen years later, Hobbes wrote that "the cause of my writing that book [i.e., *Leviathan*] was the consideration of what the ministers before, and in the beginning of, the civil war, by their preaching and writing did contribute thereunto." See Thomas Hobbes, *Six Lessons to the Professors of Mathematics*, in Sir William Molesworth, ed., *The English Works of Thomas Hobbes* (London: Longman, Brown, Green, and Longmans, 1845), p. 335.

207 *"a Civill Warr with the Pen"*: Quoted in Steven Shapin and Simon Schaffer, *Leviathan and the Air Pump* (Princeton, NJ: Princeton University Press, 1985), p. 290.

207 *Deciding which opinions and doctrines should be taught*: Hobbes, *Leviathan*, 18:9.

7. Thomas Hobbes, Geometer

212 *"made him in love with geometry"*: The account is from John Aubrey's biography, "Thomas Hobbes," p. 332.

213 *Samuel Sorbière hailed him*: Sorbière is quoted in Douglas M. Jesseph, *Squaring the Circle: The War between Hobbes and Wallis* (Chicago: University of Chicago Press, 1999), p. 6.

213 *"the only science"*: Hobbes, *Leviathan*, 4:12.

213 *"there can be no certainty of the last conclusion"*: Ibid., 5:4.

214 *"Empusa"*: All quotations in this passage are from Thomas Hobbes, "Elements of Philosophy, the First Section, Concerning Body," in Molesworth, ed., *The English Works of Thomas Hobbes*, pp. vii–xii, "The Author's Epistle Dedicatory."

214 *"no older than my own book"*: Since Hobbes published *De corpore* in 1655, and *De cive* came out in 1642, true civil philosophy is no more than thirteen years old.

214 *"revenge myself of envy by encreasing it"*: Hobbes, "Elements of Philosophy," pp. vii–xii.

215 *"For who is so stupid"*: Hobbes, *Leviathan*, 5:16.

215 *"Physics, ethics, and politics"*: Thomas Hobbes, dedicatory epistle to *De principiis et rationcinatione geometrarum* (London: Andrew Crooke, 1666), quoted in Jesseph, *Squaring the Circle*, p. 282.

215 *"fright and drive away this metaphysical Empusa"*: Hobbes, "Elements of Philosophy," pp. vii–xii.

216 *"because we make the commonwealth ourselves"*: Thomas Hobbes, *Six Lessons to the Professors of Mathematiques, One of Geometry, the Other of Astronomy, in the Chairs Set Up by the Noble and Learned Sir Henry Savile, in the University of Oxford* (London: Andrew Crooke, 1656), reprinted in Molesworth, ed., *The English Works of Thomas Hobbes*, 7:181–356. The quote is from p. 184.

216 *were as indisputably correct*: On the geometrical power of the *Leviathan*'s decrees, see also Shapin and Schaffer, *Leviathan and the Air Pump*, p. 253.

216 *"the skill of making and maintaining commonwealths"*: Hobbes, *Leviathan*, 20:19.

219 *it should have no unsolved, not to mention insoluble, problems*: On Hobbes's insistence that geometry should have no unsolved problems, see Hobbes, *De homine*, 2.10.5, quoted in Jesseph, *Squaring the Circle*, p. 221.

219 *"The quadrature of the circle"*: The account is based on Jesseph, *Squaring the Circle*, pp. 22–26; and Archimedes, "Measurement of a Circle," chap. 6 in E. J. Dijketerhuis, *Archimedes* (Princeton, NJ: Princeton University Press, 1987), pp. 222–23.

222 *the three classical problems were simply insoluble*: On views on the solubility of the quadrature of the circle, see Jesseph, *Squaring the Circle*, pp. 25–26.

222 *"that which has no parts"*: For Hobbes's discussion of Euclid's definitions, see *Six Lessons*, 7:201.

223 *"If the magnitude of a body which is moved"*: Hobbes, *De corpore*, 2.8.11, reprinted in Latin in Thomas Hobbes, *Opera philosophica* (London:

John Bohn, 1839), 1:98–99, more commonly known as Hobbes's *Opera Latina*. The passage is translated in Jesseph, *Squaring the Circle*, pp. 76–77.

223 *Points have a size, lines have a width*: On Hobbes's view that points have size and lines have width in order to construct geometrical bodies, see Hobbes, *Six Lessons*, p. 318.

224 *"conatus"*: For Hobbes's discussion of his concepts "conatus" and "impetus," see *De corpore*, 3.15.2, and *Opera Latina*, 1:177–78, both translated in Jesseph, *Squaring the Circle*, pp. 102–103.

224 *"instead of saying that a line is long"*: Sorbière, *A Voyage to England*, p. 94.

225 *"Every demonstration is flawed"*: Hobbes, *De principiis*, chap. 12, quoted in Jesseph, *Squaring the Circle*, p. 135.

225 *"speake loudest in his praise"*: Ward published this anonymously in [Seth Ward], *Vindiciae academiarum* (Oxford: L. Litchfield, 1654), p. 57, quoted in Jesseph, *Squaring the Circle*, p. 126.

225 *It was a trap, and Hobbes knew it*: On the beginnings of the war between Hobbes and Wallis, and Ward's role in it, see Jesseph, *Squaring the Circle*, p. 126.

226 *"problematically"*: The disclaimer is in Hobbes, *De corpore* (1655), p. 181. Translation from Latin in Jesseph, *Squaring the Circle*, p. 128. The titles of the chapters are in *De corpore*, p. 171.

227 *were all gleefully exposed*: See John Wallis, *Elenchus geometriae Hobbianae* (Oxford: H. Hall for John Crooke, 1655).

227 *"Solutions of Problems that have hitherto remained insoluble"*: Sorbière, *A Voyage to England*, p. 94.

227 *The attempted proofs of the quadrature*: For a modern exposition of two of these proofs, see Jesseph, "Two of Hobbes's Quadratures from *De corpore*, Part 3, Chapter 20," in Jesseph, *Squaring the Circle*, pp. 368–76.

227 *squaring of the circle is impossible*: For a fuller discussion of the impossibility of squaring the circle, see Jesseph, *Squaring the Circle*, pp. 22–28.

229 *"solved some most difficult problems"*: Hobbes's list of accomplishments is included in Aubrey, "Thomas Hobbes," pp. 400–401. Translated in Jesseph, *Squaring the Circle*, pp. 3–4.

8. Who Was John Wallis?

230 *"we* Tast *and* See *it to be so"*: John Wallis, *Truth Tried* (London: Richard Bishop for Samuel Gellibrand, 1643), pp. 60–61.

232 *"Beside his constant preaching"*: The account of Wallis's childhood and

the quotes are from John Wallis, "Autobiography," in Christoph J. Scriba, "The Autobiography of John Wallis, F.R.S.," *Notes and Records of the Royal Society of London* 25, no. 1 (June 1970): 21–23.

233 "Mr Holbech *was very kind to me*": Ibid., p. 25.

233 "*scarce bred any man that was loyall to his Prince*": The quote is from *The Autobiography of Sir John Bramston*, quoted in Vivienne Larminie, "Holbeach [Holbech], Martin," in *Oxford Dictionary of National Biography* (Oxford: Oxford University Press, 2004–12).

233 "*were scarce looked upon as* Accademical *studies*": Wallis, "Autobiography," p. 27.

235 "*a pleasing Diversion at spare hours*": Ibid.

237 "*Knowledge is no Burthen*": Ibid., p. 29.

238 *the trial of Archbishop William Laud*: Agnes Mary Clerke, "Wallis, John (1616–1703)," *Dictionary of National Biography*, vol. 59 (1899).

239 A Serious and Faithful Representation: The pamphlet's full title was *A Serious and Faithful Representation of the Judgements of Ministers of the Gospel within the Province of London, Contained in a Letter from Them to the General and His Counsel of War, Delivered to His Excellency by Some of the Subscribers, January 18, 1649* (printed in Edinburgh, 1703).

241 "*now to be my serious study*": Wallis, "Autobiography," p. 40.

241 *two mathematical treatises*: Wallis's two treatises were *De sectionibus conicis* (Oxford: Leon Lichfield, 1655) and *Arithmetica infinitorum* (Oxford: Leon Lichfield, 1656), both included in Wallis, *Opera mathematicorum* (Oxford: Leonard Lichfield, 1656–57), vol. 2.

242 Presbyterian *referred to respectable clergymen*: Wallis wrote, "When they were called *Presbyterians* it was not in the sense of *Anti-Episcopal*, but *Anti-Independents*." See Wallis, "Autobiography," p. 35.

242 *custos archivorum*: On Wallis's election as keeper of the archives and Stubbes's opposition, see Christoph J. Scriba, "John Wallis," in Gillispie, ed., *Dictionary of Scientific Biography*. On Stubbe and Hobbes, see Jesseph, *Squaring the Circle*, p. 12. On the demonization of Hobbes, see Mintz, "Thomas Hobbes," in Gillispie, ed., *Dictionary of Scientific Biography*.

242 *he be burned at the stake*: John Aubrey, "Thomas Hobbes," p. 339nc.

246 "*Their first purpose*": Thomas Sprat, *History of the Royal Society of London* (London: T.R., 1667), p. 53.

246 "*such a candid and unpassionate company as that was*": Ibid., pp. 55–56.

246 *Now, with his new companions*: On Wallis's involvement with the "Invisible College" during the Interregnum, and of the group's diverse fields of interest, see Wallis, "Autobiography," pp. 39–40.

247 "*continued such meetings in* Oxford": Wallis, "Autobiography," p. 40.

248 Philosophical Transactions: The other candidate for "first scientific journal" is *Le journal des scavans* of the French Academy of Sciences, whose first issue appeared two months before *Philosophical Transactions*.

249 *"gives us room to differ, without animosity"*: Sprat, *History of the Royal Society*, p. 56.

250 *"they* work *and* think *in company"*: Ibid., p. 427.

250 *use their experience to reconstitute the entire body politic*: For more on the early Royal Society and its mission to recast English political life and prevent a return to the disastrous dogmatism of the Interregnum, see Shapin and Schaffer, *Leviathan and the Air-Pump*; Margaret C. Jacob, *The Newtonians and the English Revolution 1689–1720* (New York: Gordon and Breach, 1990), first published in 1976; James R. Jacob, *Robert Boyle and the English Revolution* (New York: Burt Franklin and Co., 1977); Barbara J. Shapiro, *Probability and Certainty in Seventeenth Century England* (Princeton, NJ: Princeton University Press, 1983); Steven Shapin, *A Social History of Truth: Civility and Science in Seventeenth-Century England* (Chicago: University of Chicago Press, 1995).

251 *Cartesian philosophy*: The most concise summary of Descartes's philosophy is contained in his eminently readable *Discourse on the Method*, first published anonymously in Leiden in 1637 as *Discours de la méthode*.

251 *"more imperious, and impatient of contradiction"*: Sprat's views on the dangers of dogmatism can be found in Sprat, *History of the Royal Society*, p. 33.

252 *"slowness of the increase of knowledge amongst men"*: Ibid., p. 428.

252 *"The* reason *of men's contemning all* Jurisdiction *and* Power": Ibid., p. 430.

252 *"the most fruitful parent of* Sedition *is* Pride": Ibid., p. 428–29.

252 *a public statement of the goals and purpose of the Royal Society*: On Sprat and the grandees of the Royal Society, see the "Introduction" in Jackson I. Cope and Harold Whitmore Jones, eds., *The History of the Royal Society by Thomas Sprat* (St. Louis, MO: Washington University Studies, 1958), esp. pp. xiii–xiv.

252 *dogmatism leads to sedition*: In Sprat's words, "it gives them fearless confidence in their own judgments, it leads them from contending in sport to opposition in earnest . . . in the *State* as well as in the *Schools*." See Sprat, *History of the Royal Society*, p. 429.

252 *"the influence of experiments is* Obedience to the Civil Government": Sprat, *History of the Royal Society*, p. 427.

253 *"that teaches men* humility": On the beneficial effects of experimentalism, see ibid., p. 429.

253 *"one great man"*: For the Royal Society's idolization of Bacon, see ibid.,
    p. 35.
253 *Francis Bacon, Lord Chancellor to James I*: Bacon's major works include
    The Advancement of Learning (1605), *Novum organum* (1620), and *New
    Atlantis* (1627).
254 *True knowledge, worthy of*: True knowledge, in the Aristotelian scheme,
    was referred to as *scientia*, and required absolute certainty based on
    logical reasoning and ancient authority.
255 *"the daintiness and pride of mathematicians"*: The quote is from Francis
    Bacon, "Of the Dignity and Advancement of Learning," book 3, chap. 6,
    in James Spedding, ed., *The Philosophical Works of Francis Bacon*, vol. 4
    (London: Longman and Co., 1861), p. 370.
257 *It was a conundrum that left the Society with an ambivalence toward
    mathematics*: On the early Royal Society's ambivalence toward mathe-
    matics, and particularly Robert Boyle's suspicion of the field, see Shapin,
    A Social History of Truth, chap. 7.

9. Mathematics for a New World

258 *"It hath been my endeavour"*: John Wallis, "Autobiography," in Scriba,
    "The Autobiography of John Wallis, F.R.S.," p. 42.
259 *"On Conic Sections"*: John Wallis, *De sectionibus conicis, nova methodo
    expositis, tractatus* (Oxford: Leon Lichfield, 1655). On the publication
    of this treatise, along with the *Arithmetica infinitorum*, see Jacqueline
    Stedall, trans., *The Arithmetic of Infinitesimals, John Wallis, 1656* (New
    York: Springer-Verlag, 2004), p. xvii.
259 *"any plane is made up, so to speak, of infinite parallel lines"*: Wallis, *De
    sectionibus conicis*, prop. 1, in Wallis, *Opera mathematica* (Oxford: The-
    atro Sheldoniana, 1695), p. 297.
260 *"the area of the triangle is equal to the base times half the altitude"*: Wallis,
    De sectionibus conicis, prop. 3, in *Opera mathematica*, p. 299.
262 *When criticized by Fermat*: Fermat's criticisms are included in a wide-
    ranging correspondence of Wallis's *Arithmetica infinitorum*, which Wal-
    lis published in 1658 under the title *Commercium epistolicum*. Fermat's
    letters were published in French translation in volumes 2 and 3 of Paul
    Tannery and Charles Henri, eds., *Oeuvres de Fermat* (Paris: Gauthiers-
    Villars et Fils, 1894–96).
263 *"Mathematical entities exist"*: The quote is from Wallis, *Mathesis univer-
    salis* (Oxford: Leon Lichfield, 1657), chapter 3; reprinted in Wallis,
    Opera Mathematica (Oxford: Sheldonian Theatre, 1695), p.21.
264 *"it seems not in the power of the Will to reject"*: Ibid., pp. 60–61.

264 *"a general proposition may become known by induction"*: Wallis, *Arithmetica infinitorum* (Oxford: Leon Lichfield, 1656), p. 1, prop. 1. Translation is from Stedall, trans., *The Arithmetic of Infinitesimals*, p. 13.

266 *study minute creatures under a microscope*: Hooke's startling enlarged images of common insects and microbes invisible to the naked eye were published in Robert Hooke, *Micrographia or Some Physiological Descriptions of Minute Bodies Made by Magnifying Glasses* (London: John Allestry, 1667).

269 *"If there is taken a series of quantities"*: Wallis, *Arithmetica infinitorum*, prop. 2, from Stedall, trans., *The Arithmetic of Infinitesimals*, p. 14. Wallis includes an additional step demonstrating that what is true of the series 0, 1, 2, 3 . . . is also true of any arithmetic series beginning with 0.

270 *"a very good Method of Investigation"*: Wallis's discussion of induction is in John Wallis, *A Treatise of Algebra, Both Historical and Practical* (London: John Playford, 1685), p. 306.

270 *Wallis moved on to do the same for more complex series*: Wallis, *Arithmetica infinitorum*, prop. 19, from Stedall, trans., *The Arithmetic of Infinitesimals*, p. 26.

270 *"quantities that are as squares of arithmetic proportionals"*: Wallis, *Arithmetica infinitorum*, prop. 21, from Stedall, trans., *The Arithmetic of Infinitesimals*, p. 27.

271 *"quantities that are as cubes of arithmetic proportionals"*: Wallis, *Arithmetica infinitorum*, prop. 41, from Stedall, trans., *The Arithmetic of Infinitesimals*, p. 40.

272 *must be true for all powers m of natural numbers*: Wallis, *Arithmetica infinitorum*, prop. 44, from Stedall, trans., *The Arithmetic of Infinitesimals*, p. 42.

274 *he was engaged in a lively debate with Wallis*: Wallis published the entire exchange as *Commercium epistolicum de quaestionibus quibusdam mathematicis nuper habitum* (Oxford: A. Lichfield, 1658). In addition to Wallis and Fermat, it included letters from Sir Kenelm Digby, Lord Brouncker, Bernard Frénicle de Bessy, and Frans van Schooten. Fermat's critique of the *Arithmetica infinitorum* is contained mostly in "Epistola XIII," a letter from Fermat to Lord Brouncker, written in French, that was forwarded to Wallis. Fermat's contributions to the exchange are also published in Paul Tannery and Charles Henry, eds., *Oeuvres de Fermat*, vols. 2 and 3 (Paris: Gauthier-Villars et Fils, 1894 and 1896). "Epistola XIII" of Wallis's *Commercium epistolicum* is printed here as letter LXXXV in 2:347–53.

274 *"But his method of demonstration"*: Fermat to Digby, August 15, 1657, Epistola XII on p. 21 of the *Commercium epistolicum*. Also letter LXXXIV in Tannery and Henry, eds., *Oeuvres de Fermat*, 2:343.

275 *"one must settle for nothing less than a demonstration"*: "Epistola XIII," on pp. 27–28 of the *Commercium epistolicum*. Also letter LXXXV in Tannery and Henry, eds., *Oeuvres de Fermat*, 2:352.

275 *was fully answered in Cavalieri's books*: Wallis's assertion that his method is derived from Cavalieri is first stated in the dedication to the *Arithmetica infinitorum*, from Stedall, trans., *The Arithmetic of Infinitesimals*, pp. 1–2.

275 *"was already done to his hand by Cavallerius"*: Wallis, *Treatise of Algebra*, p. 305. The equivalence of Cavalieri's method of indivisibles and the method of exhaustion is discussed on p. 280, and the composition of lines, surfaces, and solids on p. 285.

275 *"You may find this work new"*: Wallis, dedication to *Arithmetica infinitorum*, from Stedall, trans., *The Arithmetic of Infinitesimals*, p. 1.

276 *"If any think them less valuable"*: Wallis, *A Treatise of Algebra*, p. 298. The claim that induction needs no additional demonstration is on p. 306.

276 *"Euclide was wont to be so pedantick"*: Quoted from Wallis, *A Treatise of Algebra*, p. 306.

276 *"[M]ost mathematicians that I have seen"*: Quoted ibid., p. 308.

276 *"a conclusive argument"*: Wallis makes the argument that the truth of a demonstration is based on the agreement of "most men" in Wallis, *A Treatise of Algebra*, pp. 307–308.

278 Elenchus geometriae Hobbianae: John Wallis, *Elenchus geometriae Hobbianae* (Oxford: H. Hall for John Crooke, 1655).

278 Decameron physiologicum: Thomas Hobbes, *Decameron physiologicum* (London: John Crooke for William Crooke, 1678).

279 *Seth Ward had traveled to London*: On Ward and Hobbes, see Jesseph, *Squaring the Circle*, p. 50.

279 *"the equal of 'Leviathan'"*: John Wallis, dedication to John Owen of *Elenchus*, folios A2r, A2v. The translation from the Latin original is from letter 37 in Peter Toon, ed., *The Correspondence of John Owen* (Cambridge: James Clarke and Co. Ltd., 1970), pp. 86–88.

280 *"I have done that business for which Dr. Wallis receives the wages"*: Thomas Hobbes, Epistle Dedicatory to Henry Lord Pierrepont to *Six Lessons*. See Molesworth, ed., *The English Works of Thomas Hobbes*, 7:185.

280 *"the most deformed necessary business which you do in your chambers"*: Hobbes, *Six Lessons*, 7:248.

280 *"scab of symbols"*: Ibid., 7:316.

280 *"the pompous ostentation of Lines and Figures"*: Wallis, *A Treatise of Algebra*, p. 298.

280 *"You do shift and wiggle"*: Thomas Hobbes, *STIGMAI, or markes of the*

absurd geometry, rural language, Scottish church-politicks, and barbarisms of John Wallis (London: Andrew Crooke, 1657), p. 12, quoted in Stedall, trans., *The Arithmetic of Infinites*, pp. xxix–xxx.

280 *"Here comes the beare to be bayted!"*: The anecdote is included in Aubrey's biography of Hobbes, "Thomas Hobbes," p. 340.

281 *"should I undertake to refute his Geometry"*: John Wallis, dedication to John Owen of *Elenchus*, p. 86.

281 *"set such store by geometry"*: John Wallis, *Elenchus*, p. 108, quoted in Jesseph, *Squaring the Circle*, p. 341.

281 *"there is no more to be feared of this Leviathan"*: John Wallis, dedication to John Owen of *Elenchus*, p. 87.

281 *"how little he understands this mathematics"*: Wallis to Huygens, January 11, 1659, quoted in Jesseph, *Squaring the Circle*, p. 70.

282 *"like Beetles from my egestions"*: Hobbes, *Six Lessons*, 7:324.

283 *"I do not wish to change, confirm, or argue"*: Hobbes to Sorbière, 7/17 March, 1664, quoted in Jesseph, *Squaring the Circle*, pp. 272–73.

283 *"Who ever, before you"*: Wallis, *Elenchus*, p. 6, quoted in Jesseph, *Squaring the Circle*, p. 78–79.

284 *"was not so much to shew a Method of Demonstrating things already known"*: Wallis, *Treatise of Algebra*, p. 305.

284 *"with all the ecclesiastics of England"*: The discussion of the motive for writing *Leviathan* is in Hobbes, *Six Lessons*, 7:335. The letter to Sorbière is quoted in Simon Schaffer, "Wallification: Thomas Hobbes on School Divinity and Experimental Pneumatics," *Studies in History and Philosophy of Science* 19 (1988): 286.

285 *"Egregious logicians and geometricians"*: Hobbes, *Six Lessons*, 7:308.

285 *"Is this the language of geometry?"*: Ibid.

285 *"If you say that by the parallels you mean infinitely little parallelograms"*: Ibid., 7:310.

286 *"it could hardly have been proposed by a sane person"*: Thomas Hobbes, *Lux mathematica* (1672), in William Molesworth, ed., *Thomae Hobbes Malmesburiensis opera philosophica*, vol. 5 (London: Longman, Brown, Green, and Longmans, 1845), p. 110, quoted and translated in Jesseph, *Squaring the Circle*, p. 182.

286 *"Nor can there be anything infinitely small"*: Hobbes, *Lux mathematica*, 5:109, quoted and translated in Jesseph, *Squaring the Circle*, p. 182.

Epilogue

289 *the dozens of sermons he delivered*: The sermons were collected in John Wallis, *Three Sermons Concerning the Sacred Trinity* (London: Thomas

Parkhurst, 1691); and John Wallis, *Theological Discourses and Sermons on Several Occasions* (London: Thomas Parkhurst, 1692).

289 *"a man of most admirable fine parts"*: The quote is by his younger contemporary, the English antiquarian Thomas Hearne. Quoted in "John Wallis," in Sidney Lee, ed., *Dictionary of National Biography*, vol. 49 (London: Smith, Elder, and Co., 1899), p. 144.

ACKNOWLEDGMENTS

The roots of this book go far back, to my first year as a graduate student at Stanford, when I wrote a paper arguing that infinitesimals were politically subversive in seventeenth-century Europe. In the following years my research interests carried me elsewhere, first to the maritime culture of early modern exploration, and then to the "romantic turn" in mathematics in the early nineteenth century. But I never forgot that early insight, and never doubted that I would one day tell the story. It took longer than I thought, but I finally did. And because I have been thinking about this topic for more than two decades, the list of those I have consulted and whose comments helped shape this book is a long one.

I would like to thank Timothy Lenoir, Peter Galison, and Moti Feingold, who commented on that paper years ago, as well as Douglas Jesseph, whose detailed critiques spurred me to refine and improve the argument. I spent hours talking about these issues with Christophe Lecuyer, Jutta Sperling, Phillip Thurtle, Josh Feinstein, and Patricia Mázon, my graduate school peers at the time. In later years my colleagues at UCLA were my sounding board, and I thank Margaret (Peg) Jacob, Mary Terrall, Ted Porter, Norton Wise, Soraya de Chadarevian, and Sharon Traweek for their insights and friendship. Carla Rita Palmerino kindly gave me access to her notes from the Jesuit Archives, and Ugo Baldini helped guide me through the maze of Jesuit sources.

Steven Vanden Broecke became a good friend during a quarter of shared office space, and contributed penetrating comments and a deep knowledge of the early modern world. Conversations with Joan Richards and Arkady Plotnitsky helped shape my thinking on mathematics and broader culture, and Mario Biagioli and Massimo Mazzotti deepened my understanding of early modern Italy and the place of mathematics in its society. Reviel Netz's "Mathematics as Literature / Mathematics as Text" workshop gave me an opportunity to test-run some of these ideas before a lively and well-informed group, and I benefited greatly from his thoughtful suggestions. Doron Zeilberger,

Michael Harris, and Jordan Ellenberg have been generous with mathematical advice, and Siegfried Zielinski has been an example of intellectual open-mindedness. Apostolos Doxiadis, in both his writings and his public outreach, showed me that mathematics, when beautifully presented, has a broad, devoted, and enthusiastic audience.

Amanda Moon of Farrar, Straus and Giroux shepherded the book through all its stages, from acquisition to publication, always providing incisive and helpful advice. Her colleagues Debra Helfand, Delia Casa, Jenna Dolan, Debra Fried, and Jennie Cohen worked diligently on all aspects of the book from copyediting to proofreading to production, turning a bland-looking electronic file into an elegantly written and beautiful final product. Dan Gerstle read every word of an early draft and made many suggestions, and Laird Gallagher brought a sharp editorial eye to later versions of the text. Both unquestionably made this a better book. Lisa Adams of the Garamond Agency was with this project from its earliest conception to its fruition, and I can truthfully say that *Infinitesimal* would never have come to be without her advice, support, and professionalism. My childhood friend Daniel Baraz has been a constant presence in my life despite living on the other side of the world. His friendship helped sustain me throughout the process.

To Bonnie, my love: thank you for being the best wife any man could wish for. Your intelligence and support are in every page of this book. My children were with me throughout the planning, writing, and production process, but they are now embarking on their own life adventures away from home. I will miss their daily presence and companionship, as well as their energy, intelligence, and creativity, and our long talks about everything from football to the *Iliad* to the art of writing. Jordan and Ella, wherever your paths may lead, my love will always follow.

INDEX

Page numbers in *italics* refer to illustrations.